The Bronze Age

Edited by Paul F. Kisak

Contents

Chapter 1

Bronze Age

For other uses, see Bronze Age (disambiguation).

The **Bronze Age** is a time period characterized by the use of bronze, proto-writing, and other early features of urban

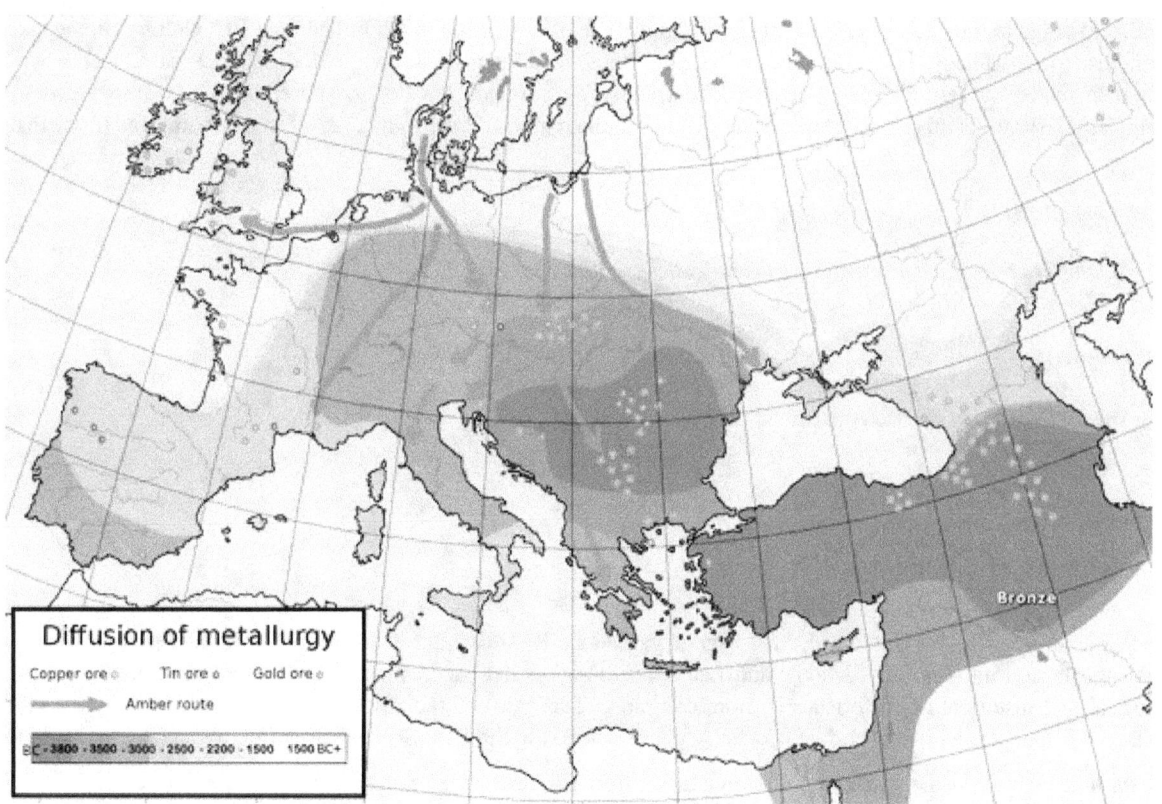

Diffusion of metallurgy in Europe and Asia Minor. The darkest areas are the oldest.

civilization. The Bronze Age is the second principal period of the three-age Stone-Bronze-Iron system, as proposed in modern times by Christian Jürgensen Thomsen, for classifying and studying ancient societies.

An ancient civilization is defined to be in the Bronze Age either by smelting its own copper and alloying with tin, arsenic, or other metals, or by trading for bronze from production areas elsewhere. Copper-tin ores are rare, as reflected in the fact that there were no tin bronzes in western Asia before trading in bronze began in the third millennium BC. Worldwide, the Bronze Age generally followed the Neolithic period, but in some parts of the world, the Copper Age served as a transition from the Neolithic to the Bronze Age. Although the Iron Age generally followed the Bronze Age, in some areas, the Iron

Age intruded directly on the Neolithic from outside the region.*[1]

Bronze Age cultures differed in their development of the first writing. According to archaeological evidence, cultures in Mesopotamia (cuneiform) and Egypt (hieroglyphs) developed the earliest viable writing systems.

1.1   History

The overall period is characterized by the full adoption of bronze in many regions, though the place and time of the introduction and development of bronze technology was not universally synchronous.*[2] Man-made tin bronze technology requires set production techniques. Tin must be mined (mainly as the tin ore cassiterite) and smelted separately, then added to molten copper to make bronze alloy. The Bronze Age was a time of extensive use of metals and of developing trade networks (See *Tin sources and trade in ancient times*).

1.1.1   Near East

Main article: Ancient Near East

The Bronze Age in the ancient Near East began with the rise of Sumer in the 4th millennium BC. Cultures in the ancient Near East (often called, "the cradle of civilization") practiced intensive year-round agriculture, developed a writing system, invented the potter's wheel, created a centralized government, law codes, and empires, and introduced social stratification, slavery, and organized warfare. Societies in the region laid the foundations for astronomy and mathematics.

Near East timeline

Dates are approximate, consult particular article for details

Age sub-divisions

The Ancient Near East Bronze Age can be divided as follows:

Mesopotamia

Main article: Ancient Mesopotamia

In Mesopotamia, the Mesopotamia Bronze Age began about 2900 BC and ended with the Kassite period. The usual tripartite division into an Early, Middle and Late Bronze Age is not used. Instead, a division primarily based on art-historical and historical characteristics is more common. The cities of the Ancient Near East housed several tens of thousands of people. Ur in the Middle Bronze Age and Babylon in the Late Bronze Age similarly had large populations.

The earliest mention of Babylonia appears on a tablet from the reign of Sargon of Akkad in the 23rd century BC. The Amorite dynasty established the city-state of Babylon in the 19th century BC. Over 100 years later, it briefly took over the other city-states and formed the first Babylonian empire during what is also called the Old Babylonian Period. Babylonia adopted the written Semitic Akkadian language for official use. By that time, the Sumerian language was no longer spoken, but was still in religious use. The Akkadian and Sumerian traditions played a major role in later Babylonian culture, and the region, even under outside rule, remained an important cultural center throughout the Bronze and Early Iron Age.

Iranian Plateau

Further information: Persian plateau

Elam was an ancient civilization located to the east of Mesopotamia. In the Old Elamite period (Middle Bronze Age), Elam consisted of kingdoms on the Iranian plateau, centered in Anshan, and from the mid-2nd millennium BC, it was centered in Susa in the Khuzestan lowlands. Its culture played a crucial role in the Gutian Empire and especially during the Achaemenid dynasty that succeeded it.

The Oxus civilization[6] was a Bronze Age Central Asian culture dated to ca. 2300–1700 BC and centered on the upper Amu Darya (Oxus). In the Early Bronze Age the culture of the Kopet Dag oases and Altyn-Depe developed a proto-urban society. This corresponds to level IV at Namazga-Depe. Altyn-Depe was a major centre even then. Pottery was wheel-turned. Grapes were grown. The height of this urban development was reached in the Middle Bronze Age c. 2300 BC, corresponding to level V at Namazga-Depe.[7] This Bronze Age culture is called the Bactria–Margiana Archaeological Complex (BMAC).

The Kulli culture,[8][9] similar to those of the Indus Valley Civilization, was located in southern Balochistan (Gedrosia) ca. 2500–2000 BC. Agriculture was the economical base of this people. At several places dams were found, providing evidence for a highly developed water management system.

Konar Sandal is associated with the hypothesized "Jiroft culture", a 3rd millennium BC culture postulated on the basis of a collection of artifacts confiscated in 2001.

Anatolia

Main article: Bronze Age Anatolia

The Hittite Empire was established in Hattusa in northern Anatolia from the 18th century BC. In the 14th century BC, the Hittite Kingdom was at its height, encompassing central Anatolia, southwestern Syria as far as Ugarit, and upper Mesopotamia. After 1180 BC, amid general turmoil in the Levant conjectured to have been associated with the sudden arrival of the Sea Peoples,[10][11] the kingdom disintegrated into several independent "Neo-Hittite" city-states, some of which survived until as late as the 8th century BC.

Arzawa in Western Anatolia during the second half of the second millennium BC likely extended along southern Anatolia in a belt that reaches from near the Turkish Lakes Region to the Aegean coast. Arzawa was the western neighbor — sometimes a rival and sometimes a vassal — of the Middle and New Hittite Kingdoms.

The Assuwa league was a confederation of states in western Anatolia that was defeated by the Hittites under an earlier Tudhaliya I, around 1400 BC. Arzawa has been associated with the much more obscure Assuwa generally located to its north. It probably bordered it, and may even be an alternative term for it (at least during some periods).

Levant

Main article: Bronze Age Levant
Further information: Canaan, Pre-history of the Southern Levant and List of archaeological periods (Levant)

In modern scholarship the chronology of the Bronze Age Levant is divided into Early/Proto Syrian; corresponding to the Early Bronze. Old Syrian; corresponding to the Middle Bronze. Middle Syrian; corresponding to the Late Bronze. The term Neo-Syria is used to designate the early Iron Age.[12]

The old Syrian period was dominated by the Eblaite first kingdom, Nagar and the Mariote second kingdom. The Akkadian conquered large areas of the Levant and were followed by the Amorite kingdoms, ca. 2000–1600 BC, which arose in Mari, Yamkhad, Qatna, Assyria,[13] From the 15th century BC onward, the term Amurru is usually applied to the region extending north of Canaan as far as Kadesh on the Orontes.

The earliest known Ugarit contact with Egypt (and the first exact dating of Ugaritic civilization) comes from a carnelian bead identified with the Middle Kingdom pharaoh Senusret I, 1971 BC–1926 BC. A stela and a statuette from the Egyptian pharaohs Senusret III and Amenemhet III have also been found. However, it is unclear at what time these monuments got to Ugarit. In the Amarna letters, messages from Ugarit ca. 1350 BC written by Ammittamru I, Niqmaddu II, and his queen, were discovered. From the 16th to the 13th century BC Ugarit remained in constant touch with Egypt and Cyprus (named Alashiya).

The Mitanni was a loosely organized state in northern Syria and south-east Anatolia from ca. 1500 BC–1300 BC. Founded by an Indo-Aryan ruling class that governed a predominately Hurrian population, Mitanni came to be a regional power after the Hittite destruction of Kassite Babylon created a power vacuum in Mesopotamia. At its beginning, Mitanni's major rival was Egypt under the Thutmosids. However, with the ascent of the Hittite empire, Mitanni and Egypt made an alliance to protect their mutual interests from the threat of Hittite domination. At the height of its power, during the 14th century BC, it had outposts centered on its capital, Washukanni, which archaeologists have located on the headwaters of the Khabur River. Eventually, Mitanni succumbed to Hittite, and later Assyrian attacks, and was reduced to a province of the Middle Assyrian Empire.

The Israelites are a Semitic people of the Ancient Near East who inhabited part of Canaan during the tribal and monarchic periods (15th to 6th centuries BC),[14][15][16][17][18] and lived in the region in smaller numbers after the fall of the monarchy. The name Israel first appears c. 1209 BC, at the end of the Late Bronze Age and the very beginning of the Iron Age, on the Merneptah Stele raised by the Egyptian Pharaoh Merneptah.

The Aramaeans are a Northwest Semitic semi-nomadic and pastoralist people who originated in what is now modern Syria (Biblical Aram) during the Late Bronze Age and the early Iron Age. Large groups migrated to Mesopotamia, where they intermingled with the native Akkadian (Assyrian and Babylonian) population. The Aramaeans never had a unified empire; they were divided into independent kingdoms all across the Near East. After the Bronze Age collapse, their political influence was confined to a number of Syro-Hittite states, which were entirely absorbed into the Neo-Assyrian Empire by the 8th century BC.

Ancient Egypt

Main article: Ancient Egypt

Early Bronze dynasties   In Ancient Egypt, the Bronze Age begins in the Protodynastic period, c. 3150 BC. The archaic *early Bronze Age of Egypt*, known as the Early Dynastic Period of Egypt,[19][20] immediately follows the unification of Lower and Upper Egypt, c. 3100 BC. It is generally taken to include the First and Second Dynasties, lasting from the Protodynastic Period of Egypt until about 2686 BC, or the beginning of the Old Kingdom. With the First Dynasty, the capital moved from Abydos to Memphis with a unified Egypt ruled by an Egyptian god-king. Abydos remained the major holy land in the south. The hallmarks of ancient Egyptian civilization, such as art, architecture and many aspects of religion, took shape during the Early Dynastic period. Memphis in the Early Bronze Age was the largest city of the time. The Old Kingdom of the regional Bronze Age[19] is the name given to the period in the 3rd millennium BC when Egypt attained its first continuous peak of civilization in complexity and achievement – the first of three "Kingdom" periods, which mark the high points of civilization in the lower Nile Valley (the others being Middle Kingdom and the New Kingdom).

The First Intermediate Period of Egypt,[21] often described as a "dark period" in ancient Egyptian history, spanned about 100 years after the end of the Old Kingdom from about 2181 to 2055 BC. Very little monumental evidence survives from this period, especially from the early part of it. The First Intermediate Period was a dynamic time when rule of Egypt was roughly divided between two competing power bases: Heracleopolis in Lower Egypt and Thebes in Upper Egypt. These two kingdoms would eventually come into conflict, with the Theban kings conquering the north, resulting in reunification of Egypt under a single ruler during the second part of the 11th Dynasty.

Middle Bronze dynasties   The Middle Kingdom of Egypt lasted from 2055 to 1650 BC. During this period, the Osiris funerary cult rose to dominate Egyptian popular religion. The period comprises two phases: the 11th Dynasty, which ruled from Thebes and the 12th[22] and 13th Dynasties centered on el-Lisht. The unified kingdom was previously considered to comprise the 11th and 12th Dynasties, but historians now at least partially consider the 13th Dynasty to belong to the Middle Kingdom.

During the Second Intermediate Period,[23] Ancient Egypt fell into disarray for a second time, between the end of the Middle Kingdom and the start of the New Kingdom. It is best known for the Hyksos, whose reign comprised the 15th and 16th dynasties. The Hyksos first appeared in Egypt during the 11th Dynasty, began their climb to power in the 13th Dynasty, and emerged from the Second Intermediate Period in control of Avaris and the Delta. By the 15th Dynasty, they ruled lower Egypt, and they were expelled at the end of the 17th Dynasty.

Late Bronze dynasties   The New Kingdom of Egypt, also referred to as the Egyptian Empire, lasted from the 16th to the 11th century BC. The New Kingdom followed the Second Intermediate Period and was succeeded by the Third Intermediate Period. It was Egypt's most prosperous time and marked the peak of Egypt's power. The later New Kingdom, i.e. the 19th and 20th Dynasties (1292–1069 BC), is also known as the Ramesside period, after the eleven pharaohs that took the name of Ramesses.

1.1.2   Central Asia

Seima-Turbino Phenomenon

Main article: Seima-Turbino Phenomenon

The Altai Mountains in what is now southern Russia and central Mongolia have been identified as the point of origin of a cultural enigma termed the Seima-Turbino Phenomenon.[24] It is conjectured that changes in climate in this region around 2000 BC and the ensuing ecological, economic and political changes triggered a rapid and massive migration westward into northeast Europe, eastward into China and southward into Vietnam and Thailand [25] across a frontier of some 4,000 miles.[24] This migration took place in just five to six generations and led to peoples from Finland in the west to Thailand in the east employing the same metal working technology and, in some areas, horse breeding and riding.[24] It is further conjectured that the same migrations spread the Uralic group of languages across Europe and Asia: some 39 languages of this group are still extant, including Hungarian, Finnish and Estonian.[24] However, recent genetic testings of sites in south Siberia and Kazakhstan (Andronovo horizon) would rather support a spreading of the bronze technology via Indo-European migrations eastwards, as this technology was well known for quite a while in western regions.[26][27]

1.1.3   East Asia

East Asia timeline

Dates are approximate, consult particular article for details

China

Further information: History of China

Historians disagree about the dates of a "Bronze Age" in China. The difficulty lies in the term "Bronze Age", as it has been applied to signify a period in history when bronze tools replaced stone tools, and, later, were themselves replaced by iron ones. The medium of the new "Age" made that of the old obsolete. In China, however, any attempt to establish a definite set of dates for a Bronze Age is complicated by two factors:

1. arrival of iron smelting technology, and

2. persistence of bronze objects.

The earliest bronze artifacts have been found in the Majiayao culture site (between 3100 and 2700 BC),[28][29] and from then on, the society gradually grew into the Bronze Age.

Bronze metallurgy in China originated in what is referred to as the Erlitou (Wade–Giles: *Erh-li-t'ou*) period, which some historians argue places it within the range of dates controlled by the Shang dynasty.[30] Others believe the Erlitou sites belong to the preceding Xia (Wade–Giles: *Hsia*) dynasty.[*][31] The U.S. National Gallery of Art defines the Chinese Bronze Age as the "period between about 2000 BC and 771 BC," a period that begins with Erlitou culture and ends abruptly with the disintegration of Western Zhou rule.[*][32] Though this provides a concise frame of reference, it overlooks the continued importance of bronze in Chinese metallurgy and culture. Since this is significantly later than the discovery of bronze in Mesopotamia, bronze technology could have been imported rather than discovered independently in China. While there may be reason to believe that bronzework developed inside China separately from outside influence,[*][33][*][34] the discovery of European mummies in Xinjiang suggests a possible route of transmission from the West.[*][35]

The Shang Dynasty[*][36][*][37] of the Yellow River Valley rose to power after the Xia Dynasty. While some direct information about the Shang Dynasty comes from Shang-era inscriptions on bronze artifacts, most comes from oracle bones — turtle shells, cattle scapulae, or other bones – which bear glyphs that form the first significant corpus of recorded Chinese characters.

Iron is found from the Zhou Dynasty, but its use is minimal. Chinese literature dating to the 6th century BC attests a knowledge of iron smelting, yet bronze continues to occupy the seat of significance in the archaeological and historical record for some time after this.[*][38] Historian W. C. White argues that iron did not supplant bronze "at any period before the end of the Zhou dynasty (256 BC)" and that bronze vessels make up the majority of metal vessels all the way through the Later Han period, or to 221 BC.[*][39] [It is unclear what White referred to: The former Han dynasty was 206-25 BC, the later Han dynasty ended 220 AD. On iron, readers may prefer to refer to this newer book: Wagner, Donald B. Iron and Steel in Ancient China. Leiden, Netherlands; New York: E.J. Brill, 1993.]

The Chinese bronze artifacts generally are either utilitarian, like spear points or adze heads, or "ritual bronzes", which are more elaborate versions in precious materials of everyday vessels, as well as tools and weapons. Examples are the numerous large sacrificial tripods known as dings in Chinese; there are many other distinct shapes. Surviving identified Chinese ritual bronzes tend to be highly decorated, often with the *taotie* motif, which involves highly stylized animal face(s). These appear in three main motif types: those of demons, of symbolic animals, and of abstract symbols.[*][40] Many large bronzes also bear cast inscriptions that are the great bulk of the surviving body of early Chinese writing and have helped historians and archaeologists piece together the history of China, especially during the Zhou Dynasty (1046-256 BC).

The bronzes of the Western Zhou Dynasty document large portions of history not found in the extant texts that were often composed by persons of varying rank and possibly even social class. Further, the medium of cast bronze lends the record they preserve a permanence not enjoyed by manuscripts.[*][41] These inscriptions can commonly be subdivided into four parts: a reference to the date and place, the naming of the event commemorated, the list of gifts given to the artisan in exchange for the bronze, and a dedication.[*][42] The relative points of reference these vessels provide have enabled historians to place most of the vessels within a certain time frame of the Western Zhou period, allowing them to trace the evolution of the vessels and the events they record.[*][43]

Korea

Main articles: Gojoseon and Mumun Pottery Period

The beginning of the Bronze Age on the peninsula is around 900 BC – 800 BC.[*][44][*][45] Although the Korean Bronze Age culture derives from the Liaoning and Manchuria, it exhibits unique typology and styles, especially in ritual objects.[*][46]

The Mumun pottery period is named after the Korean name for undecorated or plain cooking and storage vessels that form a large part of the pottery assemblage over the entire length of the period, but especially 850-550 BC. The Mumun period is known for the origins of intensive agriculture and complex societies in both the Korean Peninsula and the Japanese Archipelago.

The Middle Mumun pottery period culture of the southern Korean Peninsula gradually adopted bronze production (c. 700–600? BC) after a period when Liaoning-style bronze daggers and other bronze artifacts were exchanged as far as the interior part of the Southern Peninsula (c. 900–700 BC). The bronze daggers lent prestige and authority to the personages who wielded and were buried with them in high-status megalithic burials at south-coastal centres such as the Igeum-dong site. Bronze was an important element in ceremonies and as for mortuary offerings until 100.

See also: Liaoning bronze dagger culture

1.1.4   South Asia

South Asia timeline

Dates are approximate, consult particular article for details

Indus Valley

Main article: Indus Valley civilization

The Bronze Age on the Indian subcontinent began around 3300 BC with the beginning of the Indus Valley civilization. Inhabitants of the Indus Valley, the Harappans, developed new techniques in metallurgy and produced copper, bronze, lead and tin. The Indian Bronze Age was followed by the Iron Age Vedic Period. The Late Harappan culture, which dates from 1900 BC to 1400 BC, overlapped the transition from the Bronze Age to the Iron Age; thus it is difficult to date this transition accurately.

1.1.5   Southeast Asia

Dating back to the Neolithic Age, the first bronze drums, called the Dong Son drums, have been uncovered in and around the Red River Delta regions of Vietnam and Southern China. These relate to the prehistoric Dong Son Culture of Vietnam. In Ban Chiang, Thailand, (Southeast Asia) bronze artifacts have been discovered dating to 2100 BC.*[47] However, according to the radiocarbon dating on the human and pig bones in Ban Chiang, some scholars propose that the initial Bronze Age in Ban Chiang was in late 2nd millennium.*[48] In Nyaunggan, Burma, bronze tools have been excavated along with ceramics and stone artifacts. Dating is still currently broad (3500–500 BC).*[49] Ban Non Wat, excavated by Charles Higham, was a rich site with over 640 graves excavated that gleaned many complex bronze items that may have had social value connected to them.*[50]

Ban Chiang, however, is the most thoroughly documented site while having the clearest evidence of metallurgy when it comes to Southeast Asia. With a rough date range of late third millennium BC to the first millennium AD, this site alone has various artifacts such as burial pottery (dating from 2100 BC – 1700 BC), fragments of Bronze, copper-base bangles, and much more. What's interesting about this site, however, isn't just the old age of the artifacts but the fact that this technology suggested on-site casting from the very beginning. The on-site casting supports the theory that Bronze was first introduced in Southeast Asia as fully developed which therefore shows that Bronze was actually innovated from a different country.*[25] Some scholars believe that the copper-based metallurgy was disseminated from northwest and central China via south and southwest areas such as Guangdong province and Yunnan province and finally into southeast Asia around 1,000 BC.*[51]

Archaeological research in Northern Vietnam indicates an increase in rates of infectious disease following the advent of metallurgy; skeletal fragments in sites dating to the early and mid-Bronze Age evidence a greater proportion of lesions than in sites of earlier periods.*[52] There are a few possible implications of this. One is the increase contact with bacterial and/or fungal pathogens due to increase population density and land clearing/ cultivation. The other one is decreased levels of immunocompetence in the Metal age due to changes in diet caused by agriculture. The last is that there may have been an emergence of infectious disease in the Da But period that evolved into a more virulent form in the metal period.*[52] Archaeology also suggests that Bronze Age metallurgy may not have been as significant a catalyst in social stratification and warfare in Southeast Asia as in other regions, social distribution shifting away from chiefdom-states to a heterarchical network.*[53] Data analyses of sites such as Ban Lum Khao, Ban Na Di, Non Nok Tha, Khok Phanom Di, and Nong Nor have consistently led researchers to conclude that there was no forentrenched hierarchy.*[54]

1.1.6   Europe

Main article: Bronze Age in Europe

European timeline

A few examples of named Bronze Age cultures in Europe in roughly relative order.

Dates are approximate, consult particular article for details

The chosen cultures overlapped in time and the indicated periods do not fully correspond to their estimated extents.

Aegean

Main article: Aegean Civilization

The Aegean Bronze Age began around 3200 BC,[*][55] when civilizations first established a far-ranging trade network. This network imported tin and charcoal to Cyprus, where copper was mined and alloyed with the tin to produce bronze. Bronze objects were then exported far and wide, and supported the trade. Isotopic analysis of tin in some Mediterranean bronze artifacts points to the fact that they may have originated from Great Britain.[*][56]

Knowledge of navigation was well developed at this time, and reached a peak of skill not exceeded (except perhaps by Polynesian sailors) until 1730 when the invention of the chronometer enabled the precise determination of longitude.

The Minoan civilization based in Knossos on the island of Crete appears to have coordinated and defended its Bronze Age trade. Illyrians are also believed to have roots in the early Bronze Age. Ancient empires valued luxury goods in contrast to staple foods, leading to famine.[*][57]

Aegean Collapse   Main article: Bronze Age collapse

Bronze Age collapse theories have described aspects of the end of the Age in this region. At the end of the Bronze Age in the Aegean region, the Mycenaean administration of the regional trade empire followed the decline of Minoan primacy.[*][58] Several Minoan client states lost much of their population to famine and/or pestilence. This would indicate that the trade network may have failed, preventing the trade that would previously have relieved such famines and prevented illness caused by malnutrition. It is also known that in this era the breadbasket of the Minoan empire, the area north of the Black Sea, also suddenly lost much of its population, and thus probably some cultivation.

The Aegean Collapse has been attributed to the exhaustion of the Cyprus forests causing the end of the bronze trade.

These forests are known to have existed into later times, and experiments have shown that charcoal production on the scale necessary for the bronze production of the late Bronze Age would have exhausted them in less than fifty years.

Aegean Collapse has also been attributed to the fact that as iron tools became more common, the main justification for the tin trade ended, and that trade network ceased to function as it did formerly.[*][62] The colonies of the Minoan empire then suffered drought, famine, war, or some combination of those three, and had no access to the distant resources of an empire by which they could easily recover.

The Thera eruption occurred around the Aegean Collapse, 110 km (68 mi) north of Crete. Speculation include a tsunami from Thera (more commonly known today as Santorini) destroyed Cretan cities. A tsunami may have destroyed the Cretan navy in its home harbour, which then lost crucial naval battles; so that in the LMIB/LMII event (c. 1450 BC) the cities of Crete burned and the Mycenaean civilization took over Knossos. If the eruption occurred in the late 17th century BC (as most chronologists now think) then its immediate effects belong to the Middle to Late Bronze Age transition, and not to the end of the Late Bronze Age; but it could have triggered the instability that led to the collapse first of Knossos and then of Bronze Age society overall. One such theory looks to the role of Cretan expertise in administering the empire, post-Thera. If this expertise was concentrated in Crete, then the Mycenaeans may have made political and commercial mistakes in administering the Cretan empire.

Archaeological findings, including some on the island of Thera, suggest that the centre of Minoan Civilization at the time of the eruption was actually on Thera rather than on Crete . According to this theory, the catastrophic loss of the political, administrative and economic centre by the eruption as well as the damage wrought by the tsunami to the coastal towns and villages of Crete precipitated the decline of the Minoans. A weakened political entity with a reduced economic and military capability and fabled riches would have then been more vulnerable to human predators. Indeed, the Santorini Eruption is usually dated to c. 1630 BC, while the Mycenaean Greeks first enter the historical record a few decades later, c. 1600 BC. Thus, the later Mycenaean assaults on Crete (c.1450 BC) and Troy (c.1250 BC) are revealed as mere continuations of the steady encroachments of the Greeks upon the weakened Minoan world.

Central Europe

See also: Bronze Age in Southeastern Europe and Bronze Age in Romania

In Central Europe, the early Bronze Age Unetice culture (1800–1600 BC) includes numerous smaller groups like the Straubing, Adlerberg and Hatvan cultures. Some very rich burials, such as the one located at Leubingen with grave gifts crafted from gold, point to an increase of social stratification already present in the Unetice culture. All in all, cemeteries of this period are rare and of small size. The Unetice culture is followed by the middle Bronze Age (1600–1200 BC) Tumulus culture, which is characterised by inhumation burials in tumuli (barrows). In the eastern Hungarian Körös tributaries, the early Bronze Age first saw the introduction of the Mako culture, followed by the Otomani and Gyulavarsand cultures.

The late Bronze Age Urnfield culture, (1300–700 BC) is characterized by cremation burials. It includes the Lusatian culture in eastern Germany and Poland (1300–500 BC) that continues into the Iron Age. The Central European Bronze Age is followed by the Iron Age Hallstatt culture (700–450 BC).

Important sites include:

- Biskupin (Poland)

- Nebra (Germany)

- Vráble (Slovakia)

- Zug-Sumpf, Zug, Switzerland

The Bronze Age in Central Europe has been described in the chronological schema of German prehistorian Paul Reinecke. He described Bronze A1 (Bz A1) period (2300–2000 BC : triangular daggers, flat axes, stone wrist-guards, flint arrowheads) and Bronze A2 (Bz A2) period (1950–1700 BC : daggers with metal hilt, flanged axes, halberds, pins with perforated spherical heads, solid bracelets) and phases Hallstatt A and B (Ha A and B).

South Europe

The Apennine culture (also called Italian Bronze Age) is a technology complex of central and southern Italy spanning the Chalcolithic and Bronze Age proper. The Camuni were an ancient people of uncertain origin (according to Pliny the Elder, they were Euganei; according to Strabo, they were Rhaetians) who lived in Val Camonica – in what is now northern Lombardy – during the Iron Age, although human groups of hunters, shepherds and farmers are known to have lived in the area since the Neolithic.

Located in Sardinia and Corsica, the Nuragic civilization lasted from the early Bronze Age (18th century BC) to the 2nd century AD, when the islands were already Romanized. They take their name from the characteristic nuragic towers, which evolved from the pre-existing megalithic culture, which built dolmens and menhirs. The nuraghe towers are unanimously considered the best preserved and largest megalithic remains in Europe. Their effective use is still debated: some scholars considered them as monumental tombs, others as Houses of the Giants, other as fortresses, ovens for metal fusion, prisons or, finally, temples for a solar cult. Around the end of the third millennium BC, Sardinia exported towards Sicily a *Culture* that built small dolmens, trilithic or polygonal shaped, that served as tombs as it has been ascertained in the Sicilian dolmen of "Cava dei Servi". From this region they reached Malta island and other countries of Mediterranean basin.*[63]

The Terramare was an early Indo-European civilization in the area of what is now Pianura Padana (northern Italy) before the arrival of the Celts, and in other parts of Europe. They lived in square villages of wooden stilt houses. These villages were built on land, but generally near a stream, with roads that crossed each other at right angles. The whole complex denoted the nature of a fortified settlement. Terramare were widespread in the Pianura Padana (specially along the Panaro river, between Modena and Bologna) and in the rest of Europe. The civilization developed in the Middle and Late Bronze Age, between the 17th and the 13th centuries BC.

The Castellieri culture developed in Istria during the Middle Bronze Age. It lasted for more than a millennium, from the 15th century BC until the Roman conquest in the 3rd century BC. It takes its name from the fortified boroughs (*Castellieri*, Friulian *cjastelir*) that characterized the culture.

The Canegrate culture developed from the mid-Bronze Age (13th century BC) till the Iron Age in the Pianura Padana, in what are now western Lombardy, eastern Piedmont and Ticino. It takes its name from the township of Canegrate where, in the 20th century, some fifty tombs with ceramics and metal objects were found. The Canegrate culture migrated from the northwest part of the Alps and descended to Pianura Padana from the Swiss Alps passes and the Ticino.

The Golasecca culture developed starting from the late Bronze Age in the Po plain. It takes its name from Golasecca, a locality next to the Ticino where, in the early 19th century, abbot Giovanni Battista Giani excavated its first findings (some fifty tombs with ceramics and metal objects). Remains of the Golasecca culture span an area of c. 20,000 square kilometers south to the Alps, between the Po, Sesia and Serio rivers, dating from the 9th to the 4th century BC.

West Europe

Atlantic Bronze Age   Main article: Atlantic Bronze Age

The Atlantic Bronze Age is a cultural complex of the period of approximately 1300–700 BC that includes different cultures in Portugal, Andalusia, Galicia and the British Isles. It is marked by economic and cultural exchange. Commercial contacts extend to Denmark and the Mediterranean. The Atlantic Bronze Age was defined by a number of distinct regional centres of metal production, unified by a regular maritime exchange of some of their products.

Great Britain   Main article: Bronze Age Britain

In Great Britain, the Bronze Age is considered to have been the period from around 2100 to 750 BC. Migration brought new people to the islands from the continent. Recent tooth enamel isotope research on bodies found in early Bronze Age graves around Stonehenge indicate that at least some of the migrants came from the area of modern Switzerland. The Beaker culture displayed different behaviours from the earlier Neolithic people, and cultural change was significant. Integration is thought to have been peaceful, as many of the early henge sites were seemingly adopted by the newcomers. The rich Wessex culture developed in southern Britain at this time. Additionally, the climate was deteriorating; where once the weather was warm and dry it became much wetter as the Bronze Age continued, forcing the population away from easily defended sites in the hills and into the fertile valleys. Large livestock farms developed in the lowlands and appear to have contributed to economic growth and inspired increasing forest clearances. The Deverel-Rimbury culture began to emerge in the second half of the Middle Bronze Age (c. 1400–1100 BC) to exploit these conditions. Devon and Cornwall were major sources of tin for much of western Europe and copper was extracted from sites such as the Great Orme mine in northern Wales. Social groups appear to have been tribal but with growing complexity and hierarchies becoming apparent.

Burial of dead (which, until this period, had usually been communal) became more individual. For example, whereas in the Neolithic a large chambered cairn or long barrow housed the dead, Early Bronze Age people buried their dead in individual barrows (also commonly known and marked on modern British Ordnance Survey maps as tumuli), or sometimes in cists covered with cairns.

The greatest quantities of bronze objects in England were discovered in East Cambridgeshire, where the most important finds were recovered in Isleham (more than 6500 pieces).*[64] Alloying of copper with zinc or tin to make brass or bronze was practised soon after the discovery of copper itself. One copper mine at Great Orme in North Wales, extended to a depth of 70 meters.*[65] At Alderley Edge in Cheshire, carbon dates have established mining at around 2280 to 1890 BC (at 95% probability).*[66] The earliest identified metalworking site (Sigwells, Somerset) is much later, dated by Globular Urn style pottery to approximately the 12th century BC. The identifiable sherds from over 500 mould fragments included a perfect fit of the hilt of a sword in the Wilburton style held in Somerset County Museum.*[67]

Ireland   See also: Atlantic Bronze Age

The Bronze Age in Ireland commenced around 2000 BC, when copper was alloyed with tin and used to manufacture Ballybeg type flat axes and associated metalwork. The preceding period is known as the Copper Age and is characterised by the production of flat axes, daggers, halberds and awls in copper. The period is divided into three phases: Early Bronze Age (2000–1500 BC), Middle Bronze Age (1500–1200 BC), and Late Bronze Age (1200 – c. 500 BC). Ireland is also known for a relatively large number of Early Bronze Age burials.

One of the characteristic types of artifact of the Early Bronze Age in Ireland is the flat axe. There are five main types of flat axes: Lough Ravel (c. 2200 BC), Ballybeg (c. 2000 BC), Killaha (c. 2000 BC), Ballyvalley (c. 2000–1600 BC), Derryniggin (c. 1600 BC), and a number of metal ingots in the shape of axes.*[68]

North Europe

Main article: Nordic Bronze Age

The Bronze Age in Northern Europe spans the entire 2nd millennium BC (Unetice culture, Urnfield culture, Tumulus culture, Terramare culture, Lusatian culture) lasting until c. 600 BC. The Northern Bronze Age was both a period and a Bronze Age culture in Scandinavian pre-history, c. 1700–500 BC, with sites that reached as far east as Estonia. Succeeding the Late Neolithic culture, its ethnic and linguistic affinities are unknown in the absence of written sources. It is followed by the Pre-Roman Iron Age.

Even though Northern European Bronze Age cultures were relatively late, and came in existence via trade, sites present rich and well-preserved objects made of wool, wood and imported Central European bronze and gold. Many rock carvings depict ships, and the large stone burial monuments known as stone ships suggest that shipping played an important role. Thousands of rock carvings depict ships, most probably representing sewn plank built canoes for warfare, fishing and trade. These may have a history as far back as the neolithic period and continue into the Pre-Roman Iron Age, as shown by the Hjortspring boat. There are many mounds and rock carving sites from the period. Numerous artifacts of bronze and gold are found. No written language existed in the Nordic countries during the Bronze Age. The rock carvings have been dated through comparison with depicted artifacts.

Caucasus

Arsenical bronze artifacts of the Maykop culture in the North Caucasus have been dated around the 4th millennium BC.*[69] This innovation resulted in the circulation of arsenical bronze technology over southern and eastern Europe.*[70]

Pontic–Caspian steppe

The Yamna culture is a late copper age/early Bronze Age culture of the Southern Bug/Dniester/Ural region (the Pontic steppe), dating to the 36th–23rd centuries BC. The name also appears in English as Pit Grave Culture or Ochre Grave Culture. The Catacomb culture, c. 2800–2200 BC, is several related early Bronze Age cultures occupying what is presently Ukraine. The Srubna culture was a Late Bronze Age (18th–12th centuries BC) culture. It is a successor to the Yamna culture, the Pit Grave culture and the Poltavka culture.

1.1.7   Americas

See also: Metallurgy in pre-Columbian America

The Moche civilization of South America independently discovered and developed bronze smelting.*[71] Bronze technology was developed further by the Incas and used widely both for utilitarian objects and sculpture.*[72] Later appearance of limited bronze smelting in West Mexico (see Metallurgy in pre-Columbian Mesoamerica) suggests either contact of that region with Andean cultures or separate discovery of the technology. The Calchaquí people of Northwest Argentina had a Bronze technology.*[73]

1.2   Outside the Bronze Age

1.2.1   Japan

Main article: Jōmon period

The Jōmon period lasted until 300 BC and, towards the end of the period, the Japanese archipelago experienced the introduction of bronze and iron simultaneously. Bronze and iron smelting techniques spread to the Japanese archipelago through immigration and trade from the Korean peninsula and the Chinese mainland. Iron was mainly used for agricultural and other tools, whereas ritual and ceremonial artifacts were mainly made of bronze. Formerly, scholarly theories suggested that a bronze and iron using Yamato people gradually spread across the Japanese archipelago, conquering and assimilating the Jōmon people and their descendants, as well as pushing them east and north. Current archaeology suggests a more complex picture of the "Jōmon-Yayoi transition," including as regards ethnic categories; see the article on Yayoi people.

1.2.2   Africa

See also: Prehistoric North Africa

Although North Africa was influenced to a certain extent by European Bronze Age cultures (for example, traces of the Bell beaker tradition are found in Morocco), it has long been believed that Africa did not have its own metallurgy traditions until the Phoenician colonization (ca. 1100 BC) of North Africa, and that it remained attached to the Neolithic way of life. The civilization of Ancient Egypt, whose influence did not extensively cover Africa outside of the Nile's reach, was believed to be the sole exception to this rule as regards the whole range of ancient cultures of Africa. Recently, however, some discoveries have been made that appear to contradict these views.

In the Termit region of eastern Niger, its ancient inhabitants are now thought to have become the first iron smelting people in West Africa and among the first in the world at around 1500 BC. Iron and copper working then continued to spread southward to Nigeria, and then moved elsewhere in the continent, reaching South Africa around AD 200. The widespread use of iron revolutionized the Bantu-speaking farming communities who adopted it, driving out and absorbing the rock tool using hunter-gatherer societies they encountered as they expanded to farm wider areas of savannah. The technologically superior Bantu-speakers spread across southern Africa and became wealthy and powerful, producing iron for tools and weapons in large, industrial quantities.

1.3   See also

- Middle Bronze Age migrations (Ancient Near East)

- Namazga V and Altyndepe

- Oxhide ingot

- Synoptic table of the principal old world prehistoric cultures

1.3.1 Seafaring

- Dover Bronze Age Boat—the earliest known seagoing plank-built vessel

- Ferriby Boats

- Langdon Bay hoard—see also Dover Museum

1.4 Notes

[1] Iron In Africa: Revising The History : Unesco. Portal.unesco.org. Retrieved on 2013-07-28.

[2] Bronze was independently discovered in the Maykop culture of the North Caucasus as early as the mid-4th millennium BC, which makes them the producers of the oldest known bronze. However, the Maykop culture only had arsenical bronze, a naturally occurring alloy. Other regions developed bronze and its associated technology at different periods.

[3] The Near East period dates and phases are unrelated to the bronze chronology of other regions of the world.

[4] Piotr Bienkowski, Alan Ralph Millard (editors). *Dictionary of the ancient Near East.* Page 60.

[5] Amélie Kuhr. *The Ancient Near East, c. 3000–330 BC.* Page 9.

[6] Dalton, O. M., Franks, A. W., & Read, C. H. (1905). The treasure of the Oxus: With other objects from ancient Persia and India. London: British Museum.

[7] V.M. Masson, The Bronze Age in Khorasan and Transoxiana, chapter 10 in A.H. Dani and Vadim Mikhaĭlovich Masson (eds.), History of civilizations of Central Asia, volume 1: The dawn of civilization: earliest times to 700 BC

[8] Possehl, G. L. (1986). Kulli: An exploration of ancient civilization in Asia. Durham, N.C: Carolina Academic Press

[9] Piggott, S. (1961). Prehistoric India to 1000 B.C. Baltimore: Penguin Book.

[10] Killebrew, Ann E. (2013), "The Philistines and Other "Sea Peoples" in Text and Archaeology", *Society of Biblical Literature Archaeology and biblical studies* (Society of Biblical Lit) **15**: 2, ISBN 978-1-58983-721-8. Quote: "First coined in 1881 by the French Egyptologist G. Maspero (1896), the somewhat misleading term "Sea Peoples" encompasses the ethnonyms Lukka, Sherden, Shekelesh, Teresh, Eqwesh, Denyen, Sikil / Tjekker, Weshesh, and Peleset (Philistines). [Footnote: The modern term "Sea Peoples" refers to peoples that appear in several New Kingdom Egyptian texts as originating from "islands" (tables 1-2; Adams and Cohen, this volume; see, e.g., Drews 1993, 57 for a summary). The use of quotation marks in association with the term "Sea Peoples" in our title is intended to draw attention to the problematic nature of this commonly used term. It is noteworthy that the designation "of the sea" appears only in relation to the Sherden, Shekelesh, and Eqwesh. Subsequently, this term was applied somewhat indiscriminately to several additional ethnonyms, including the Philistines, who are portrayed in their earliest appearance as invaders from the north during the reigns of Merenptah and Ramesses Ill (see, e.g., Sandars 1978; Redford 1992, 243, n. 14; for a recent review of the primary and secondary literature, see Woudhuizen 2006). Henceforce the term Sea Peoples will appear without quotation marks.]"

[11] The End of the Bronze Age: Changes in Warfare and the Catastrophe Ca. 1200 B.C., Robert Drews, p48–61 Quote: "The thesis that a great "migration of the Sea Peoples" occurred ca. 1200 B.C. is supposedly based on Egyptian inscriptions, one from the reign of Merneptah and another from the reign of Ramesses III. Yet in the inscriptions themselves such a migration nowhere appears. After reviewing what the Egyptian texts have to say about 'the sea peoples', one Egyptologist (Wolfgang Helck) recently remarked that although some things are unclear, "eins ist aber sicher: Nach den agyptischen Texten haben wir es nicht mit einer 'Volkerwanderung' zu tun." Thus the migration hypothesis is based not on the inscriptions themselves but on their interpretation."

[12] Mogens Herman Hansen (2000). *A Comparative Study of Thirty City-state Cultures: An Investigation, Volume 21.* p. 57.

[13] under Shamshi-Adad I

[14] Finkelstein, Israel. "Ethnicity and origin of the Iron I settlers in the Highlands of Canaan: Can the real Israel stand up?." The Biblical archaeologist 59.4 (1996): 198-212.

[15] Finkelstein, Israel. The archaeology of the Israelite settlement. Jerusalem: Israel Exploration Society, 1988.

[16] Finkelstein, Israel, and Nadav Na'aman, eds. From nomadism to monarchy: archaeological and historical aspects of early Israel. Yad Izhak Ben-Zvi, 1994.

[17] Finkelstein, Israel. "The archaeology of the United Monarchy: an alternative view." Levant 28.1 (1996): 177-187.

[18] Finkelstein, Israel, and Neil Asher Silberman. The Bible Unearthed: Archaeology's New Vision of Ancient Israel and the Origin of Sacred Texts. Simon and Schuster, 2002.

[19] Karin Sowada and Peter Grave. Egypt in the Eastern Mediterranean during the Old Kingdom.

[20] Lukas de Blois and R. J. van der Spek. An Introduction to the Ancient World. Page 14.

[21] Hansen, M. H. (2000). A comparative study of thirty city-state cultures: An investigation conducted by the Copenhagen Polis Centre. Copenhagen: Det Kongelike Danske Videnskabernes Selskab. Page 68.

[22] Othmar Keel and Christoph Uehlinger. *Gods, goddesses, and images of God in ancient Israel*, 1998. Page 17 (cf. "The first phase (Middle Bronze Age IIA) runs roughly parallel to the Egyptian Twelfth Dynasty")

[23] Bruce G. Trigger. *Ancient Egypt: a social history*. 1983. Page 137. (cf. ... "for the Middle Kingdom and Second Intermediate Period it is the Middle Bronze Age" .)

[24] Keys, David (January 2009). "Scholars crack the code of an ancient enigma" . *BBC History Magazine* **10** (1): 9.

[25] White, Joyce; Hamilton, Elizabeth (2009). "The Transmission of Early Bronze Technology to Thailand: New Perspectives" . *Journal of World Prehistory* **22**: 357–397. doi:10.1007/s10963-009-9029-z.

[26] C. Lalueza-Fox et al. 2004. *Unravelling migrations in the steppe: mitochondrial DNA sequences from ancient central Asians*

[27] C. Keyser et al. 2009. Ancient DNA provides new insights into the history of south Siberian Kurgan people. Human Genetics.

[28] Martini, I. Peter (2010). *Landscapes and Societies: Selected Cases*. Springer. p. 310. ISBN 90-481-9412-1.

[29] Higham, Charles (2004). *Encyclopedia of ancient Asian civilizations*. Infobase Publishing. p. 200. ISBN 0-8160-4640-9.

[30] Chang, K. C.: "Studies of Shang Archaeology" , pp. 6–7, 1. Yale University Press, 1982.

[31] Chang, K. C.: "Studies of Shang Archaeology" , p. 1. Yale University Press, 1982.

[32] "Teaching Chinese Archaeology, Part Two —NGA" . Nga.gov. Retrieved 2010-01-17.

[33] Li-Liu; The Chinese Neolithic, Cambridge University Press, 2005

[34] *Shang and Zhou Dynasties: The Bronze Age of China* Heilbrunn Timeline Retrieved May 13, 2010

[35] Jan Romgard (2008). "Questions of Ancient Human Settlements in Xinjiang and the Early Silk Road Trade, with an Overview of the Silk Road Research Institutions and Scholars in Beijing, Gansu, and Xinjiang" (PDF). *Sino-Platonic Papers* (185).

[36] Also known as the Yin Dynasty.

[37] Thorp, R. L. (2005). China in the early bronze age: Shang civilization. Philadelphia: Univ. of Pennsylvania Press.

[38] Barnard, N.: "Bronze Casting and Bronze Alloys in Ancient China" , p. 14. The Australian National University and Monumenta Serica, 1961.

[39] White, W. C.: "Bronze Culture of Ancient China" , p. 208. University of Toronto Press, 1956.

[40] Erdberg, E.: "Ancient Chinese Bronzes" , p. 20. Siebenbad-Verlag, 1993.

[41] Shaughnessy, E. L.: "Sources of Western Zhou History" , pp. xv–xvi. University of California Press, 1982.

[42] Shaughnessy, E. L. "Sources of Western Zhou History" , pp. 76–83. University of California Press, 1982.

[43] Shaughnessy, E. L. "Sources of Western Zhou History" , p. 107

[44] Carter J. Eckert, el., "Korea, Old and New: History", 1990, pp. 9

[45] "1000 BC to 300 AD: Korea | Asia for Educators | Columbia University". Afe.easia.columbia.edu. Retrieved 2012-08-03.

[46]

[47] "Bronze from Ban Chiang, Thailand: A view from the Laboratory" (PDF). Museum.upenn.edu. Retrieved 2010-01-17.

[48] Higham, C., Higham, T., Ciarla, R., Douka, K., Kijngam, A., & Rispoli, F. (2011). The Origins of the Bronze Age of Southeast Asia. Journal of world prehistory, 24(4), 227-274.

[49] "Nyaunggan City —Archaeological Sites in Myanmar". Myanmartravelinformation.com. Retrieved 2010-01-17.

[50] Higham, C. F. W. (2011). The Bronze Age of Southeast Asia: New insight on social change from Ban Non Wat. Cambridge Archaeological Journal, 21(3), 365-389.

[51] Higham, C., Higham, T., Ciarla, R., Douka, K., Kijngam, A., & Rispoli, F. (2011). The Origins of the Bronze Age of Southeast Asia. Journal of world prehistory, 24(4), 227-274.

[52] Oxenham, M.F.; Thuy, N.K.; Cuong, N.L. (2005). "Skeletal evidence for the emergence of infectious disease in bronze and iron age northern Vietnam". *American Journal of Physical Anthropology* **126** (4): 359–376. doi:10.1002/ajpa.20048.

[53] White, J.C. (1995). "Incorporating Heterarchy into Theory on Socio-political Development: The Case from Southeast Asia". *Archaeological Papers of the American Anthropological Association* **6** (1): 101–123. doi:10.1525/ap3a.1995.6.1.101.

[54] O' Reilly, D.J.W. 2003. Further evidence of heterarchy in Bronze Age Thailand. Current Anthropology 44:300-306.

[55] "Ancient Greece". British Museum. Retrieved 2012-08-03.

[56] Carl Waldman, Catherine Mason. *Encyclopedia of European peoples: Volume 1*. 2006. Page 524.

[57] Lancaster, H. O. (1990). Expectations of life: A study in the demography, statistics, and history of world mortality. New York: Springer-Verlag. Page 228.

[58] Drews, R. (1993). The end of the Bronze Age: Changes in warfare and the catastrophe ca. 1200 B.C. Princeton, N.J: Princeton University Press

[59] Cities on the Sea., Swiny, S., Hohlfelder, R. L., & Swiny, H. W. (1998). Res maritimae: Cyprus and the eastern Mediterranean from prehistory to late antiquity : proceedings of the Second International Symposium "Cities on the Sea", Nicosia, Cyprus, October 18–22, 1994. Atlanta, Ga: Scholars Press.

[60] Creevey, B. (1994). The forest resources of Bronze Age Cyprus

[61] A. Bernard Knapp, Steve O. Held and Sturt W. Manning. The prehistory of Cyprus: Problems and prospects.

[62] Lockard, Craig A. (2009). Societies, Networks, and Transitions: To 600. Wadsworth Pub Co. Page 96.

[63] Piccolo, Salvatore, *op. cit.*, pp. 1 onwards.

[64] Hall and Coles, p. 81–88.

[65] O'Brien, W. (1997). *Bronze Age Copper Mining in Britain and Ireland*. Shire Publications Ltd. ISBN 0-7478-0321-8.

[66] Timberlake, S. and Prag A.J.N.W. (2005). *The Archaeology of Alderley Edge:Survey, excavation and experiment in an ancient mining landscape*. Oxford: John and Erica Hedges Ltd. p. 396.

[67] Tabor, Richard (2008). *Cadbury Castle: A hillfort and landscapes*. Stroud: The History Press. pp. 61–69. ISBN 978-0-7524-4715-5.

[68] Waddell; Eogan.

[69] Philip L. Kohl. The making of bronze age Eurasia. Page 58.

[70] Gimbutas, "The Beginning of the Bronze Age in Europe and the Indo- Europeans 3500–2500 BC," Journal of Indo-European Studies 1 (1973): 177.

[71] "El bronce y el horizonte medio". *lablaa.org*.

[72] Antonio Gutierrez. "Inca Metallurgy". Incas.homestead.com. Retrieved 2010-01-17.

[73] Ambrosetti, El bronze de la región calchaquí, Buenos Aires, 1904., accessed 28 March 2015.

1.5 References

- Figueiredo, Elin (2010) "Smelting and Recycling Evidences from the Late Bronze Age habitat site of Baioes," Journal of Archaeological Science, Volume 37, Issue 7, p. 1623–1634

- Eogan, George (1983) *The hoards of the Irish later Bronze Age*, Dublin: University College, 331p., ISBN 0-901120-77-4

- Hall, David and Coles, John (1994) *Fenland survey : an essay in landscape and persistence*, Archaeological report **1**, London : English Heritage, 170 p., ISBN 1-85074-477-7

- Pernicka, E., Eibner, C., Öztunah, Ö., Wagener, G.A. (2003) "Early Bronze Age Metallurgy in the Northeast Aegean", In: Wagner, G.A., Pernicka, E. and Uerpmann, H-P. (eds), *Troia and the Troad: scientific approaches*, Natural science in archaeology, Berlin; London : Springer, ISBN 3-540-43711-8, p. 143–172

- Piccolo, Salvatore (2013). *Ancient Stones: The Prehistoric Dolmens of Sicily*. Abingdon (GB): Brazen Head Publishing, ISBN 978-09565106-2-4,

- Waddell, John (1998) *The prehistoric archaeology of Ireland*, Galway University Press, 433 p., ISBN 1-901421-10-4

- Siklosy et al. (2009): Bronze Age volcanic event recorded in stalagmites by combined isotope and trace element studies. Rapid Communications in Mass Spectrometry, 23/6, 801-808. doi:10.1002/rcm.3943

- Roberts, B.W., Thornton, C.P. and Pigott, V.C. 2009. Development of Metallurgy in Eurasia. Antiquity 83, 112-122.

1.6 Further reading

- Childe, V. G. (1930). *The bronze age*. New York: The Macmillan Company.

- Fong, Wen (ed.) (1980). *The great bronze age of China: an exhibition from the People's Republic of China*. New York: The Metropolitan Museum of Art. ISBN 0-87099-226-0.

- Kelleher, Bradford (1980). *Treasures from the Bronze Age of China: An exhibition from the People's Republic of China, the Metropolitan Museum of Art, New York*. New York: Ballantine Books. ISBN 0-87099-230-9.

- Wagner, Donald B. (1993). *Iron and Steel in Ancient China*. Leiden, Netherlands; New York: E.J. Brill.

- Kuijpers, M. H. G. (2008). *Bronze Age metalworking in the Netherlands (c. 2000-800 BC): A research into the preservation of metallurgy related artefacts and the social position of the smith*. Leiden: Sidestone Press.

- Müller-Lyer, F. C.; Lake, E. C.; Lake, H. A. (1921). *The history of social development*. New York: Alfred A. Knopf.

- Pittman, Holly (1984). *Art of the Bronze Age: southeastern Iran, western Central Asia, and the Indus Valley*. New York: The Metropolitan Museum of Art. ISBN 978-0-87099-365-7.

- Higham, C. F. W. (2011). The Bronze Age of Southeast Asia: New insight on social change from Ban Non Wat. Cambridge Archaeological Journal, 21(3), 365-389.

1.7 External links

- Bronze Age Experimental Archeology and Museum Reproductions

- Umha Aois – Reconstructed Bronze Age metal casting

- Umha Aois – ancient bronze casting videoclip

- Reconstructing the Danish Trundholm Sun Chariot

- Ancient bronze idol 13 Cent B.C.: Northern Russia (Russian)

- Aegean and Balkan Prehistory articles, site-reports and bibliography database concerning the Aegean, Balkans and Western Anatolia

- Li; et al. (2010). "Evidence that a West-East admixed population lived in the Tarim Basin as early as the early Bronze Age" (PDF). *BMC Biology* **8**: 15. doi:10.1186/1741-7007-8-15.

- "The Transmission of Early Bronze Technology to Thailand: New Perspectives"

1.7.1 Seafaring

- Divers unearth Bronze Age hoard off the coast of Devon

Chapter 2

Three-age system

For other uses of Three Ages, see Three Ages (disambiguation).

The **three-age system** in archaeology and physical anthropology is the periodization of human prehistory and history

Trundholm sun chariot, National Museum of Denmark

into three consecutive time periods, named for their respective tool-making technologies:

- The Stone Age
- The Bronze Age
- The Iron Age

Jomon pottery, Japanese Stone Age.

2.1 Origin

The concept of dividing pre-historical ages into systems based on metals extends far back in European history, but the present archaeological system of the three main ages—stone, bronze and iron—originates with the Danish archaeologist Christian Jürgensen Thomsen (1788–1865), who placed the system on a more scientific basis by typological and chronological studies, at first, of tools and other artifacts present in the Museum of Northern Antiquities in Copenhagen (later the National Museum of Denmark). He later used artifacts and the excavation reports published or sent to him by Danish archaeologists who were doing controlled excavations. His position as curator of the museum gave him enough visibility to become highly influential on Danish archaeology. A well-known and well-liked figure, he explained his system in person to visitors at the museum, many of them professional archaeologists.

2.1.1   The Metallic Ages of Hesiod

In his poem, *Works and Days*, the ancient Greek poet Hesiod possibly between 750 and 650 BC, defined five successive Ages of Man: 1. Golden, 2. Silver, 3. Bronze, 4. Heroic and 5. Iron.*[1] Only the Bronze Age and the Iron Age are based on the use of metal:*[2]

> "... then Zeus the father created the third generation of mortals, the age of bronze ... They were terrible and strong, and the ghastly action of Ares was theirs, and violence. ... The weapons of these men were bronze, of bronze their houses, and they worked as bronzesmiths. There was not yet any black iron."

Hesiod knew from the traditional poetry, such as the Iliad, and the heirloom bronze artifacts that abounded in Greek society, that before the use of iron to make tools and weapons, bronze had been the preferred material and iron was not smelted at all. He did not continue the manufacturing metaphor, but mixed his metaphors, switching over to the market value of each metal. Iron was cheaper than bronze, so there must have been a golden and a silver age. He portrays a sequence of metallic ages, but it is a degradation rather than a progression. Each age has less of a moral value than the preceding. Of his own age he says:*[3] "And I wish that I were not any part of the fifth generation of men, but had died before it came, or had been born afterward."

2.1.2   The Progress of Lucretius

The moral metaphor of the ages of metals continued. Lucretius, however, replaced moral degradation with the concept of progress,*[4] which he conceived to be like the growth of an individual human being. The concept is evolutionary:*[5]

> "For the nature of the world as a whole is altered by age. Everything must pass through successive phases. Nothing remains forever what it was. Everything is on the move. Everything is transformed by nature and forced into new paths ... The Earth passes through successive phases, so that it can no longer bear what it could, and it can now what it could not before."

The Romans believed that the species of animals, including man, were spontaneously generated from the materials of the Earth, because of which the Latin word *mater*, "mother," descends to English-speakers as matter and material. In Lucretius the Earth is a mother, Venus, to whom the poem is dedicated in the first few lines. She brought forth mankind by spontaneous generation, a view that, removed to the molecular stage, and stripped of its anthropomorphism, is the same as in today's biological chemistry. Having been given birth as a species, man must grow to maturity by analogy with individual men. The different phases of their collective life are marked by the accumulation of customs to form material civilization:*[6]

> "The earliest weapons were hands, nails and teeth. Next came stones and branches wrenched from trees, and fire and flame as soon as these were discovered. Then men learnt to use tough iron and copper. With copper they tilled the soil. With copper they whipped up the clashing waves of war, ... Then by slow degrees the iron sword came to the fore; the bronze sickle fell into disrepute; the ploughman began to cleave the earth with iron, ..."

Lucretius envisioned a pre-technological man that was "far tougher than the men of today ... They lived out their lives in the fashion of wild beasts roaming at large." *[7] The next stage was the use of huts, fire, clothing, language and the family. City-states, kings and citadels followed them. Lucretius supposes that the initial smelting of metal occurred accidentally in forest fires. The use of copper followed the use of stones and branches and preceded the use of iron.

2.1.3   Early lithic analysis by Michele Mercati

By the 16th century, a tradition had developed based on observational incidents, true or false, that the black objects found widely scattered in large quantities over Europe had fallen from the sky during thunderstorms and were therefore to be considered generated by lightning. They were so published by Konrad Gessner in *De rerum fossilium, lapidum et gemmarum maxime figuris & similitudinibus* at Zurich in 1565 and by many others less famous.[8] The name ceraunia, "thunderstones," had been assigned.

Ceraunia were collected by many persons over the centuries including Michele Mercati, Superintendent of the Vatican Botanical Garden in the late 16th century. He brought his collection of fossils and stones to the Vatican, where he studied them at leisure, compiling the results in a manuscript, which was published posthumously by the Vatican at Rome in 1717 as *Metallotheca*. Mercati was interested in Ceraunia cuneata, "wedge-shaped thunderstones," which seemed to him to be most like axes and arrowheads, which he now called ceraunia vulgaris, "folk thunderstones," distinguishing his view from the popular one.[9] His view was based on what may be the first in-depth lithic analysis of the objects in his collection, which led him to believe that they are artifacts and to suggest that the historical evolution of these artifacts followed a scheme.

Mercati examining the surfaces of the ceraunia noted that the stones were of flint and that they had been chipped all over by another stone to achieve by percussion their current forms. The protrusion at the bottom he identified as the attachment point of a haft. Concluding that these objects were not ceraunia he compared collections to determine exactly what they were. Vatican collections included artifacts from the New World of exactly the shapes of the supposed ceraunia. The reports of the explorers had identified them to be implements and weapons or parts of them.[10]

Mercati posed the question to himself, why would anyone prefer to manufacture artifacts of stone rather than of metal, a superior material?[11] His answer was that metallurgy was unknown at that time. He cited Biblical passages to prove that in Biblical times stone was the first material used. He also revived the 3-age system of Lucretius, which described a succession of periods based on the use of stone (and wood), bronze and iron respectively. Due to lateness of publication, Mercati's ideas were already being developed independently; however, his writing served as a further stimulus.

2.1.4   The usages of Mahudel and de Jussieu

On November 12, 1734, Nicholas Mahudel, physician, antiquarian and numismatist, read a paper at a public sitting of the Académie Royale des Inscriptions et Belles-Lettres in which he defined three "usages" of stone, bronze and iron in a chronological sequence. He had presented the paper several times that year but it was rejected until the November revision was finally accepted and published by the Academy in 1740. It was entitled *Les Monumens les plus anciens de l'industrie des hommes, et des Arts reconnus dans les Pierres de Foudres.*[12] It expanded the concepts of Antoine de Jussieu, who had gotten a paper accepted in 1723 entitled *De l'Origine et des usages de la Pierre de Foudre.*[13] In Mahudel, there is not just one usage for stone, but two more, one each for bronze and iron.

He begins his treatise with descriptions and classifications of the *Pierres de Tonnerre et de Foudre*, the ceraunia of contemporaneous European interest. After cautioning the audience that natural and man-made objects are often easily confused, he asserts that the specific "*figures*" or "formes that can be distinguished (*formes qui les font distingues*)" of the stones were man-made, not natural:[14]

> "It was Man's hand that made them serve as instruments (*C'est la main des hommes qui les leur a données pour servir d'instrumens...*)"

Their cause, he asserts, is "the industry of our forefathers (*l'industrie de nos premiers pères*)." He adds later that bronze and iron implements imitate the uses of the stone ones, suggesting a replacement of stone with metals. Mahudel is careful not to take credit for the idea of a succession of usages in time but states: "it is Michel Mercatus, physician of Clement VIII who first had this idea".[15] He does not coin a term for ages, but speaks only of the times of usages. His use of *l'industrie* foreshadows the 20th century "industries," but where the moderns mean specific tool traditions, Mahudel meant only the art of working stone and metal in general.

2.1.5   The three-age system of C. J. Thomsen

An important step in the development of the Three-age System came when the Danish antiquarian Christian Jürgensen Thomsen was able to use the Danish national collection of antiquities and the records of their finds as well as reports from contemporaneous excavations to provide a solid empirical basis for the system. He showed that artifacts could be classified into types and that these types varied over time in ways that correlated with the predominance of stone, bronze or iron implements and weapons. In this way he turned the Three-age System from being an evolutionary scheme based on intuition and general knowledge into a system of relative chronology supported by archaeological evidence. Initially, the three-age system as it was developed by Thomsen and his contemporaries in Scandinavia, such as Sven Nilsson and J.J.A. Worsaae, was grafted onto the traditional biblical chronology. But, during the 1830s they achieved independence from textual chronologies and relied mainly on typology and stratigraphy.*[17]

In 1816 Thomsen at age 27 was appointed to succeed the retiring Rasmus Nyerup as Secretary of the *Kongelige Commission for Oldsagers Opbevarung*[18] ("Royal Commission for the Preservation of Antiquities"), which had been founded in 1807.*[19] The post was unsalaried. Thomsen had independent means. At his appointment Bishop Münter said that he was an "amateur with a great range of accomplishments." Between 1816 and 1819 he reorganized the commission's collection of antiquities. In 1819 he opened the first Museum of Northern Antiquities, in Copenhagen, in a former monastery, to house the collections.*[20] It later became the National Museum.

Like the other antiquarians Thomsen undoubtedly knew of the three-age model of prehistory through the works of Lucretius, the Dane Vedel Simonsen, Montfaucon and Mahudel. Sorting the material in the collection chronologically*[21] he mapped out which kinds of artifacts co-occurred in deposits and which did not, as this arrangement would allow him to discern any trends that were exclusive to certain periods. In this way he discovered that stone tools did not co-occur with bronze or iron in the earliest deposits while subsequently bronze did not co-occur with iron - so that three periods could be defined by their available materials, stone, bronze and iron.

To Thomsen the find circumstances were the key to dating. In 1821 he wrote in a letter to fellow prehistorian Schröder:*[22]

> "nothing is more important than to point out that hitherto we have not paid enough attention to what was found together."

and in 1822:

> "we still do not know enough about most of the antiquities either; ... only future archaeologists may be able to decide, but they will never be able to do so if they do not observe what things are found together and our collections are not brought to a greater degree of perfection."

This analysis emphasizing co-occurrence and systematic attention to archaeological context allowed Thomsen to build a chronological framework of the materials in the collection and to classify new finds in relation to the established chronology, even without much knowledge of their provenience. In this way, Thomsen's system was a true chronological system rather than an evolutionary or technological system.*[23] Exactly when his chronology was reasonably well established is not clear, but by 1825 visitors to the museum were being instructed in his methods.*[24] In that year also he wrote to J.G.G. Büsching:*[25]

> "To put artifacts in their proper context I consider it most important to pay attention to the chronological sequence, and I believe that the old idea of first stone, then copper, and finally iron, appears to be ever more firmly established as far as Scandinavia is concerned."

By 1831 Thomsen was so certain of the utility of his methods that he circulated a pamphlet, "*Scandinavian Artifacts and Their Preservation*, advising archaeologists to "observe the greatest care" to note the context of each artifact. The pamphlet had an immediate effect. Results reported to him confirmed the universality of the Three-age System. Thomsen also published in 1832 and 1833 articles in the *Nordisk Tidsskrift for Oldkyndighed*, "Scandinavian Journal of Archaeology." *[26] He already had an international reputation when in 1836 the Royal Society of Northern Antiquaries published his illustrated contribution to "Guide to Scandinavian Archaeology" in which he put forth his chronology together with comments about typology and stratigraphy.

Thomsen was the first to perceive typologies of grave goods, grave types, methods of burial, pottery and decorative motifs, and to assign these types to layers found in excavation. His published and personal advice to Danish archaeologists concerning the best methods of excavation produced immediate results that not only verified his system empirically but placed Denmark in the forefront of European archaeology for at least a generation. He became a national authority when C.C Rafn,[*][27] secretary of the *Kongelige Nordiske Oldskriftselskab* ("Royal Society of Northern Antiquaries"), published his principal manuscript[*][21] in *Ledetraad til Nordisk Oldkyndighed* ("Guide to Scandinavian Archaeology")[*][28] in 1836. The system has since been expanded by further subdivision of each era, and refined through further archaeological and anthropological finds.

2.2 Stone Age subdivisions

2.2.1 The savagery and civilization of Sir John Lubbock

It was to be a full generation before British archaeology caught up with the Danish. When it did, the leading figure was another multi-talented man of independent means: John Lubbock, 1st Baron Avebury. After reviewing the Three-age System from Lucretius to Thomsen, Lubbock improved it and took it to another level, that of cultural anthropology. Thomsen had been concerned with techniques of archaeological classification. Lubbock found correlations with the customs of savages and civilization.

In his 1865 book, *Prehistoric Times*, Lubbock divided the Stone Age in Europe, and possibly nearer Asia and Africa, into the Palaeolithic and the Neolithic:[*][29]

1. "That of the Drift... This we may call the 'Palaeolithic' Period."
2. "The later, or polished Stone Age ... in which, however, we find no trace ... of any metal, excepting gold, ... This we may call the 'Neolithic' Period."
3. "The Bronze Age, in which bronze was used for arms and cutting instruments of all kinds."
4. "The Iron Age, in which that metal had superseded bronze."

By "drift" Lubbock meant river-drift, the alluvium deposited by a river. For the interpretation of Palaeolithic artifacts, Lubbock, pointing out that the times are beyond the reach of history and tradition, suggests an analogy, which was adopted by the anthropologists. Just as the paleontologist uses modern elephants to help reconstruct fossil pachyderms, so the archaeologist is justified in using the customs of the "non-metallic savages" of today to understand "the early races which inhabited our continent." [*][30] He devotes three chapters to this approach, covering the "modern savages" of the Indian and Pacific Oceans and the Western Hemisphere, but something of a deficit in what would be called today his professionalism reveals a field yet in its infancy:[*][31]

"Perhaps it will be thought ... I have selected ... the passages most unfavorable to savages. ... In reality the very reverse in the case. ... Their real condition is even worse and more abject than that which I have endeavoured to depict."

2.2.2 The elusive Mesolithic of Hodder Westropp

Sir John Lubbock's use of the terms Palaeolithic ("Old Stone Age") and Neolithic ("New Stone Age") were immediately popular. They were applied, however, in two different senses: geologic and anthropologic. In 1867-1868 Ernst Haeckel in 20 public lectures in Jena, entitled *General Morphology*, to be published in 1870, referred to the Archaeolithic, the Palaeolithic, the Mesolithic and the Caenolithic as periods in geologic history.[*][32] He could only have gotten these terms from Hodder Westropp, who took Palaeolithic from Lubbock, innovated Mesolithic ("Middle Stone Age") and Caenolithic instead of Lubbock's Neolithic. None of these terms appear anywhere, including the writings of Haeckel, before 1865. Haeckel's use was innovative.

Westropp innovated the Mesolithic and the Caenolithic in 1865, almost immediately after the publication of Lubbock's first edition. He read a paper on the topic before the Anthropological Society of London in 1865, published in 1866 in the *Memoirs*. After asserting:[*][33]

"Man, in all ages and in all stages of his development, is a tool-making animal."

Westropp goes on to define "different epochs of flint, stone, bronze or iron; ..." He never did distinguish the flint from the stone age (having realized they were one and the same), but he divided the Stone Age as follows:[34]

1. "The flint implements of the gravel-drift"

2. "The flint implements found in Ireland and Denmark"

3. "Polished stone implements"

These three ages were named respectively the Palaeolithic, the Mesolithic and the Kainolithic. He was careful to qualify these by stating:[35]

"Their presence is thus not always an evidence of a high antiquity, but of an early and barbarous state; ..."

Lubbock's savagery was now Westropp's barbarism. A fuller exposition of the Mesolithic waited for his book, *Pre-Historic Phases*, dedicated to Sir John Lubbock, published in 1872. At that time he restored Lubbock's Neolithic and defined a Stone Age divided into three phases and five stages.

The First Stage, "Implements of the Gravel Drift," contains implements that were "roughly knocked into shape."[36] His illustrations show Mode 1 and Mode 2 stone tools, basically Acheulean handaxes. Today they are in the Lower Palaeolithic.

The Second Stage, "Flint Flakes" are of the "simplest form" and were struck off cores.[37] Westropp differs in this definition from the modern, as Mode 2 contains flakes for scrapers and similar tools. His illustrations, however, show Modes 3 and 4, of the Middle and Upper Palaeolithic. His extensive lithic analysis leaves no doubt. They are, however part of Westropp's Mesolithic.

The Third Stage, "a more advanced stage" in which "flint flakes were carefully chipped into shape," produced small arrowheads from shattering a piece of flint into "a hundred pieces", selecting the most suitable and working it with a punch.[38] The illustrations show that he had microliths, or Mode 5 tools in mind. His Mesolithic is therefore partly the same as the modern.

The Fourth Stage is a part of the Neolithic that is transitional to the Fifth Stage: axes with ground edges leading to implements totally ground and polished. Westropp's agriculture is removed to the Bronze Age, while his Neolithic is pastoral. The Mesolithic is reserved to hunters.

2.2.3 Piette finds the Mesolithic

In that same year, 1872, Sir John Evans produced a massive work, *The Ancient Stone Implements*, in which he in effect repudiated the Mesolithic, making a point to ignore it, denying it by name in later editions. He wrote:[39]

"Sir John Lubbock has proposed to call them the Archaeolithic, or Palaeolithic, and the Neolithic Periods respectively, terms which have met with almost general acceptance, and of which I shall avail myself in the course of this work."

Evans did not, however, follow Lubbock's general trend, which was typological classification. He chose instead to use type of find site as the main criterion, following Lubbock's descriptive terms, such as tools of the drift. Lubbock had identified drift sites as containing Palaeolithic material. Evans added to them the cave sites. Opposed to drift and cave were the surface sites, where chipped and ground tools often occurred in unlayered contexts. Evans decided he had no choice but to assign them all to the most recent. He therefore consigned them to the Neolithic and used the term "Surface Period" for it.

Having read Westropp, Sir John knew perfectly well that all the former's Mesolithic implements were surface finds. He used his prestige to quell the concept of Mesolithic as best he could, but the public could see that his methods were not typological. The less prestigious scientists publishing in the smaller journals continued to look for a Mesolithic. For example, Isaac Taylor in *The Origin of the Aryans*, 1889, mentions the Mesolithic but briefly, asserting, however, that it formed "a transition between the Palaeolithic and Neolithic Periods." *[40] Nevertheless, Sir John fought on, opposing the Mesolithic by name as late as the 1897 edition of his work.

Meanwhile, Haeckel had totally abandoned the geologic uses of the -lithic terms. The concepts of Palaeozoic, Mesozoic and Cenozoic had been innovated in the early 19th century and were gradually becoming coin of the geologic realm. Realizing he was out of step, Haeckel started to transition to the -zoic system as early as 1876 in *The history of creation*, placing the -zoic form in parentheses next to the -lithic form.*[41]

The gauntlet was officially thrown down before Sir John by J. Allen Brown, speaking for the opposition before the Anthropological Institute on March 8, 1892. In the journal he opens the attack by striking at a "hiatus" in the record:*[42]

> "It has been generally assumed that a break occurred between the period during which ... the continent of Europe was inhabited by Palaeolithic Man and his Neolithic successor ... No physical cause, no adequate reasons have ever been assigned for such a hiatus in human existence ...."

The main hiatus at that time was between British and French archaeology, as the latter had already discovered the gap 20 years earlier and had already considered three answers and arrived at one solution, the modern. Whether Brown did not know or was pretending not to know is unclear. In 1872, the very year of Evans' publication, Mortillet had presented the gap to the Congrès international d'Anthropologie at Brussels:*[43]

> "Between the Palaeolithic and Neolithic, there is a wide and deep gap, a large hiatus."

Apparently prehistoric man was hunting big game with stone tools one year and farming with domestic animals and ground stone tools the next. Mortillet postulated a "time then unknown (*époque alors inconnue*)" to fill the gap. The hunt for the "unknown" was on. On April 16, 1874, Mortillet retracted.*[44] "That hiatus is not real (*Cet hiatus n'est pas réel*)," he said before the *Société d'Anthropologie*, asserting that it was an informational gap only. The other theory had been a gap in nature, that, because of the ice age, man had retreated from Europe. The information must now be found. In 1895 Piette stated that he had heard Lartet speak of "the remains from the intermediate period (*les vestiges de l'époque intermédiaire*)", which were yet to be discovered, but Lartet had not published this view.*[43] The gap had become a transition. However, asserted Piette:*[45]

> "I was fortunate to discover the remains of that unknown time which separated the Magdalenian age from that of polished stone axes ... it was, at Mas-d'Azil in 1887 and 1888 when I made this discovery."

He had excavated the type site of the Azilian Culture, the basis of today's Mesolithic. He found it sandwiched between the Magdalenian and the Neolithic. The tools were like those of the Danish kitchen-middens, termed the Surface Period by Evans, which were the basis of Westropp's Mesolithic. They were Mode 5 stone tools, or microliths. He mentions neither Westropp nor the Mesolithic, however. For him this was a "solution of continuity (*solution de continuité*)" To it he assigns the semi-domestication of dog, horse, cow, etc., which "greatly facilitated the work of neolithic man (*a beaucoup facilité la tàche de l'homme néolithique*)." Brown in 1892 does not mention Mas-d'Azil. He refers to the "transition or 'Mesolithic' forms" but to him these are "rough hewn axes chipped over the entire surface" mentioned by Evans as the earliest of the Neolithic.*[46] Where Piette believed he had discovered something new, Brown wanted to break out known tools considered Neolithic.

2.2.4 The Epipaleolithic and Protoneolithic of Stjerna and Obermaier

Sir John Evans never changed his mind, giving rise to a dichotomous view of the Mesolithic and a multiplication of confusing terms. On the continent, all seemed settled: there was a distinct Mesolithic with its own tools and both tools and customs were transitional to the Neolithic. Then in 1910, the Swedish archaeologist, Knut Stjerna, addressed another problem of the Three-Age System: although a culture was predominantly classified as one period, it might contain material that was the same as or like that of another. His example was the Gallery grave Period of Scandinavia. It was not uniformly Neolithic, but contained some objects of bronze and more importantly to him three different subcultures.[47]

One of these "civilisations" (sub-cultures) located in the north and east of Scandinavia[48] was rather different, featuring but few gallery graves, using instead stone-lined pit graves containing implements of bone, such as harpoon and javelin heads. He observed that they "persisted during the recent Paleolithic period and also during the Protoneolithic." Here he had used a new term, "Protoneolithic", which was according to him to be applied to the Danish kitchen-middens.[49]

Stjerna also said that the eastern culture "is attached to the Paleolithic civilization (*se trouve rattachée à la civilisation paléolithique*)." However, it was not intermediary and of its intermediates he said "we cannot discuss them here (*nous ne pouvons pas examiner ici*)." This "attached" and non-transitional culture he chose to call the Epipaleolithic, defining it as follows:[50]

> "With Epipaleolithic I mean the period during the early days that followed the age of the reindeer, the one that retained Paleolithic customs. This period has two stages in Scandinavia, that of Maglemose and that of Kunda. (*Par époque épipaléolithique j'entends la période qui, pendant les premiers temps qui ont suivi l'âge du Renne, conserve les coutumes paléolithiques. Cette période présente deux étapes en Scandinavie, celle de Maglemose et de Kunda.*)"

There is no mention of any Mesolithic, but the material he described had been previously connected with the Mesolithic. Whether or not Stjerna intended his Protoneolithic and Epipaleolithic as a replacement for the Mesolithic is not clear, but Hugo Obermaier, a German archaeologist who taught and worked for many years in Spain, to whom the concepts are often erroneously attributed, used them to mount an attack on the entire concept of Mesolithic. He presented his views in *El Hombre fósil*, 1916, which was translated into English in 1924. Viewing the Epipaleolithic and the Protoneolithic as a "transition" and an "interim" he affirmed that they were not any sort of "transformation:"[51]

> "But in my opinion this term is not justified, as it would be if these phases presented a natural evolutionary development – a progressive transformation from Paleolithic to Neolithic. In reality, the final phase of the Capsian, the Tardenoisian, the Azilian and the northern Maglemose industries are the posthumous descendants of the Palaeolithic ..."

The ideas of Stjerna and Obermaier introduced a certain ambiguity into the terminology, which subsequent archaeologists found and find confusing. Epipaleolithic and Protoneolithic cover the same cultures, more or less, as does the Mesolithic. Publications on the Stone Age after 1916 include some sort of explanation of this ambiguity, leaving room for different views. Strictly speaking the Epipaleolithic is the earlier part of the Mesolithic. Some identify it with the Mesolithic. To others it is an Upper Paleolithic transition to the Mesolithic. The exact use in any context depends on the archaeological tradition or the judgement of individual archaeologists. The issue continues.

2.2.5   Lower, middle and upper from Haeckel to Sollas

The post-Darwinian approach to the naming of periods in earth history focused at first on the lapse of time: early (Palaeo-), middle (Meso-) and late (Ceno-). This conceptualization automatically imposes a three-age subdivision to any period, which is predominant in modern archaeology: Early, Middle and Late Bronze Age; Early, Middle and Late Minoan, etc. The criterion is whether the objects in question look simple or are elaborative. If a horizon contains objects that are post-late and simpler-than-late they are sub-, as in Submycenaean.

Haeckel's presentations are from a different point of view. His *History of Creation* of 1870 presents the ages as "Strata of the Earth's Crust," in which he prefers "upper", "mid-" and "lower" based on the order in which one encounters the layers. His analysis features an Upper and Lower Pliocene as well as an Upper and Lower Diluvial (his term for the Pleistocene).[41] Haeckel, however, was relying heavily on Lyell. In the 1833 edition of *Principles of Geology* (the first) Lyell devised the terms Eocene, Miocene and Pliocene to mean periods of which the "strata" contained some (Eo-, "early"), lesser (Mio-) and greater (Plio-) numbers of "living Mollusca represented among fossil assemblages of western Europe." [52] The Eocene was given Lower, Middle, Upper; the Miocene a Lower and Upper; and the Pliocene an Older and Newer, which scheme would indicate an equivalence between Lower and Older, and Upper and Newer.

In a French version, *Nouveaux Éléments de Géologie*, in 1839 Lyell called the Older Pliocene the Pliocene and the Newer Pliocene the Pleistocene (Pleist-, "most"). Then in *Antiquity of Man* in 1863 he reverted to his previous scheme, adding "Post-Tertiary" and "Post-Pliocene." In 1873 the Fourth Edition of *Antiquity of Man* restores Pleistocene and identifies it with Post-Pliocene. As this work was posthumous, no more was heard from Lyell. Living or deceased, his work was immensely popular among scientists and laymen alike. "Pleistocene" caught on immediately; it is entirely possible that he restored it by popular demand. In 1880 Dawkins published *The Three Pleistocene Strata* containing a new manifesto for British archaeology:[53]

> "The continuity between geology, prehistoric archaeology and history is so direct that it is impossible to picture early man in this country without using the results of all these three sciences."

He intends to use archaeology and geology to "draw aside the veil" covering the situations of the peoples mentioned in proto-historic documents, such as Caesar's *Commentaries* and the *Agricola* of Tacitus. Adopting Lyell's scheme of the Tertiary, he divides Pleistocene into Early, Mid- and Late.[54] Only the Palaeolithic falls into the Pleistocene; the Neolithic is in the "Prehistoric Period" subsequent.[55] Dawkins defines what was to become the Upper, Middle and Lower Paleolithic, except that he calls them the "Upper Cave-Earth and Breccia," [56] the "Middle Cave-Earth," [57] and the "Lower Red Sand," [58] with reference to the names of the layers. The next year, 1881, Geikie solidified the terminology into Upper and Lower Palaeolithic:[59]

> "In Kent's Cave the implements obtained from the lower stages were of a much ruder description than the various objects detected in the upper cave-earth ... And a very long time must have elapsed between the formation of the lower and upper Palaeolithic beds in that cave."

The Middle Paleolithic in the modern sense made its appearance in 1911 in the 1st edition of William Johnson Sollas' *Ancient Hunters*.[60] It had been used in varying senses before then. Sollas associates the period with the Mousterian technology and the relevant modern people with the Tasmanians. In the 2nd edition of 1915 he has changed his mind for reasons that are not clear. The Mousterian has been moved to the Lower Paleolithic and the people changed to the Australian aborigines; furthermore, the association has been made with Neanderthals and the Levalloisian added. Sollas says wistfully that they are in "the very middle of the Palaeolithic epoch." Whatever his reasons, the public would have none of it. From 1911 on, Mousterian was Middle Paleolithic, except for holdouts. Alfred L. Kroeber in 1920, *Three essays on the antiquity and races of man,* reverting to Lower Paleolithic, explains that he is following Louis Laurent Gabriel de Mortillet. The English-speaking public remained with Middle Paleolithic.

2.2.6 Early and late from Worsaae through the three-stage African system

Thomsen had formalized the Three-age System by the time of its publication in 1836. The next step forward was the formalization of the Palaeolithic and Neolithic by Sir John Lubbock in 1865. Between these two times Denmark held the lead in archaeology, especially because of the work of Thomsen's at first junior associate and then successor, Jens Jacob Asmussen Worsaae, rising in the last year of his life to Kultus Minister of Denmark. Lubbock offers full tribute and credit to him in *Prehistoric Times*.

Worsaae in 1862 in *Om Tvedelingen af Steenalderen*, previewed in English even before its publication by *The Gentleman's Magazine*, concerned about changes in typology during each period, proposed a bipartite division of each age:[61]

> "Both for Bronze and Stone it was now evident that a few hundred years would not suffice. In fact, good grounds existed for dividing each of these periods into two, if not more."

He called them earlier or later. The three ages became six periods. The British seized on the concept immediately. Worsaae's earlier and later became Lubbock's palaeo- and neo- in 1865, but alternatively English speakers used Earlier and Later Stone Age, as did Lyell's 1883 edition of *Principles of Geology*, with older and younger as synonyms. As there is no room for a middle between the comparative adjectives, they were later modified to early and late. The scheme created a problem for further bipartite subdivisions, which would have resulted in such terms as early early stone age, but that terminology was avoided by adoption of Geikie's upper and lower Paleolithic.

Amongst African archaeologists, the terms Old Stone Age, Middle Stone Age and Late Stone Age are preferred.

2.2.7   Wallace's grand revolution recycled

When Sir John Lubbock was doing the preliminary work for his 1865 *magnum opus*, Charles Darwin and Alfred Russel Wallace were jointly publishing their first papers On the Tendency of Species to form Varieties; and on the Perpetuation of Varieties and Species by Natural Means of Selection. Darwins's On the Origin of Species came out in 1859, but he did not elucidate the theory of evolution as it applies to man until the Descent of Man in 1871. Meanwhile, Wallace read a paper in 1864 to the Anthropological Society of London that was a major influence on Sir John, publishing in the very next year.[62] He quoted Wallace:[63]

> From the moment when the first skin was used as a covering, when the first rude spear was formed to assist in the chase, the first seed sown or shoot planted, a grand revolution was effected in nature, a revolution which in all the previous ages of the world's history had had no parallel, for a being had arisen who was no longer necessarily subject to change with the changing universe,—a being who was in some degree superior to nature, inasmuch as he knew how to control and regulate her action, and could keep himself in harmony with her, not by a change in body, but by an advance in mind.

Wallace distinguishing between mind and body was asserting that natural selection shaped the form of man only until the appearance of mind; after then, it played no part. Mind formed modern man, meaning that result of mind, culture. Its appearance overthrew the laws of nature. Wallace used the term "grand revolution." Although Lubbock believed that Wallace had gone too far in that direction he did adopt a theory of evolution combined with the revolution of culture. Neither Wallace not Lubbock offered any explanation of how the revolution came about, or felt that they had to offer one. Revolution is an acceptance that in the continuous evolution of objects and events sharp and inexplicable disconformities do occur, as in geology. And so it is not surprising that in the 1874 Stockholm meeting of the International Congress of Anthropology and Prehistoric Archaeology, in response to Ernst Hamy's denial of any "break" between Paleolithic and Neolithic based on material from dolmens near Paris "showing a continuity between the paleolithic and neolithic folks," Edouard Desor, geologist and archaeologist, replied:[64] "that the introduction of domesticated animals was a complete revolution and enables us to separate the two epochs completely."

A revolution as defined by Wallace and adopted by Lubbock is a change of regime, or rules. If man was the new rule-setter through culture then the initiation of each of Lubbock's four periods might be regarded as a change of rules and therefore as a distinct revolution, and so Chambers's Journal, a reference work, in 1879 portrayed each of them as:[65]

> "...an advance in knowledge and civilization which amounted to a revolution in the then existing manners and customs of the world."

Because of the controversy over Westropp's Mesolithic and Mortillet's Gap beginning in 1872 archaeological attention focused mainly on the revolution at the Palaeolithic—Neolithic boundary as an explanation of the gap. For a few decades the Neolithic Period, as it was called, was described as a kind of revolution. In the 1890s, a standard term, the Neolithic Revolution, began to appear in encyclopedias such as Pears. In 1925 the Cambridge Ancient History reported:[66]

> "There are quite a large number of archaeologists who justifiably consider the period of the Late Stone Age to be a neolithic revolution and an economic revolution at the same time. For that is the period when primitive agriculture developed and cattle breeding began."

2.2.8   Vere Gordon Childe's revolution for the masses

In 1936 a champion came forward who would advance the Neolithic Revolution into the mainstream view: Vere Gordon Childe. After giving the Neolithic Revolution scant mention in his first notable work, the 1928 edition of *New Light on the Most Ancient East*, Childe made a major presentation in the first edition of *Man Makes Himself* in 1936 developing Wallace's and Lubbock's theme of the human revolution against the supremacy of nature and supplying detail on two revolutions, the Paleolithic—Neolithic and the Neolithic-Bronze Age, which he called the Second or Urban revolution.

Lubbock had been as much of an ethnologist as an archaeologist. The founders of cultural anthropology, such as Tylor and Morgan, were to follow his lead on that. Lubbock created such concepts as savages and barbarians based on the customs of then modern tribesmen and made the presumption that the terms can be applied without serious inaccuracy to the men of the Paleolithic and the Neolithic. Childe broke with this view:*[67]

> "The assumption that any savage tribe today is primitive, in the sense that its culture faithfully reflects that of much more ancient men is gratuitous."

Childe concentrated on the inferences to be made from the artifacts:*[68]

> "But when the tools ... are considered ... in their totality, they may reveal much more. They disclose not only the level of technical skill ... but also their economy .... The archaeologists's ages correspond roughly to economic stages. Each new "age" is ushered in by an economic revolution ...."

The archaeological periods were indications of economic ones:*[69]

> "Archaeologists can define a period when it was apparently the sole economy, the sole organization of production ruling anywhere on the earth's surface."

These periods could be used to supplement historical ones where history was not available. He reaffirmed Lubbock's view that the Paleolithic was an age of food gathering and the Neolithic an age of food production. He took a stand on the question of the Mesolithic identifying it with the Epipaleolithic. The Mesolithic was to him "a mere continuance of the Old Stone Age mode of life" between the end of the Pleistocene and the start of the Neolithic.*[70] Lubbock's terms "savagery" and "barbarism" do not much appear in *Man Makes Himself* but the sequel, *What Happened in History* (1942), reuses them (attributing them to Morgan, who got them from Lubbock) with an economic significance: savagery for food-gathering and barbarism for Neolithic food production. Civilization begins with the urban revolution of the Bronze Age.*[71]

2.2.9   The Pre-pottery Neolithic of Garstang and Kenyon at Jericho

Even as Childe was developing this revolution theme the ground was sinking under him. Lubbock did not find any pottery associated with the Paleolithic, asserting of its to him last period, the Reindeer, "no fragments of metal or pottery have yet been found." *[72] He did not generalize but others did not hesitate to do so. The next year, 1866, Dawkins proclaimed of Neolithic people that "these invented the use of pottery...." *[73] From then until the 1930s pottery was considered a sine qua non of the Neolithic. The term Pre-Pottery Age came into use in the late 19th century but it meant Paleolithic.

Meanwhile, the Palestine Exploration Fund founded in 1865 completing its survey of excavatable sites in Palestine in 1880 began excavating in 1890 at the site of ancient Lachish near Jerusalem, the first of a series planned under the licensing system of the Ottoman Empire. Under their auspices in 1908 Ernst Sellin and Carl Watzinger began excavation at Jericho previously excavated for the first time by Sir Charles Warren in 1868. They discovered a Neolithic and Bronze Age city there. Subsequent excavations in the region by them and others turned up other walled cities that appear to have preceded the Bronze Age urbanization.

All excavation ceased for World War I. When it was over the Ottoman Empire was no longer a factor there. In 1919 the new British School of Archaeology in Jerusalem assumed archaeological operations in Palestine. John Garstang finally resumed excavation at Jericho 1930-1936. The renewed dig uncovered another 3000 years of prehistory that was in the Neolithic but did not make use of pottery. He called it the Pre-pottery Neolithic, as opposed to the Pottery Neolithic, subsequently often called the Aceramic or Pre-ceramic and Ceramic Neolithic.

Kathleen Kenyon was a young photographer then with a natural talent for archaeology. Solving a number of dating problems she soon advanced to the forefront of British archaeology through skill and judgement. In World War II she served as a commander in the Red Cross. In 1952–58 she took over operations at Jericho as the Director of the British School, verifying and expanding Garstang's work and conclusions.*[74] There were two Pre-pottery Neolithic periods, she concluded, A and B. Moreover, the PPN had been discovered at most of the major Neolithic sites in the near East and Greece. By this time her personal stature in archaeology was at least equal to that of V. Gordon Childe. While the three-age system was being attributed to Childe in popular fame, Kenyon became gratuitously the discoverer of the PPN. More significantly the question of revolution or evolution of the Neolithic was increasingly being brought before the professional archaeologists.

2.3   Bronze age subdivisions

Danish archaeology took the lead in defining the Bronze Age, with little of the controversy surrounding the Stone Age. British archaeologists patterned their own excavations after those of the Danish, which they followed avidly in the media. References to the Bronze Age in British excavation reports began in the 1820s contemporaneously with the new system being promulgated by C.J. Thomsen. Mention of the Early and Late Bronze Age began in the 1860s following the bipartite definitions of Worsaae.

2.3.1   The tripartite system of Sir John Evans

In 1874 at the Stockholm meeting of the International Congress of Anthropology and Prehistoric Archaeology, a suggestion was made by A. Bertrand that no distinct age of bronze had existed, that the bronze artifacts discovered were really part of the Iron Age. Hans Hildebrand in refutation pointed to two Bronze Ages and a transitional period in Scandinavia. John Evans denied any defect of continuity between the two and asserted there were three Bronze Ages, "the early, middle and late bronze age." [75]

His view for the Stone Age, following Lubbock, was quite different, denying, in the *The Ancient Stone Implements*, any concept of a Middle Stone Age. In his 1881 parallel work, *The Ancient Bronze Implements*, he affirmed and further defined the three periods, strangely enough recusing himself from his previous terminology, Early, Middle and Late Bronze Age (the current forms) in favor of "an earlier and later stage" [76] and "middle" . [77] He uses Bronze Age, Bronze Period, Bronze-using Period and Bronze Civilization interchangeably. Apparently Evans was sensitive of what had gone before, retaining the terminology of the bipartite system while proposing a tripartite one. After stating a catalogue of types of bronze implements he defines his system:[78]

> "The Bronze Age of Britain may, therefore, be regarded as an aggregate of three stages: the first, that characterized by the flat or slightly flanged celts, and the knife-daggers ... the second, that characterized by the more heavy dagger-blades and the flanged celts and tanged spear-heads or daggers, ... and the third, by palstaves and socketed celts and the many forms of tools and weapons, ... It is in this third stage that the bronze sword and the true socketed spear-head first make their advent."

2.3.2   From Evans' gratuitous Copper Age to the mythical chalcolithic

In chapter 1 of his work, Evans proposes for the first time a transitional Copper Age between the Neolithic and the Bronze Age. He adduces evidence from far-flung places such as China and the Americas to show that the smelting of copper universally preceded alloying with tin to make bronze. He does not know how to classify this fourth age. On the one hand he distinguishes it from the Bronze Age. On the other hand, he includes it:[79]

> "In thus speaking of a bronze-using period I by no means wish to exclude the possible use of copper unalloyed with tin."

Evans goes into considerable detail tracing references to the metals in classical literature: Latin *aer, aeris* and Greek *chalkós* first for "copper" and then for "bronze." He does not mention the adjective of *aes*, which is *aēneus*, nor is he interested in formulating New Latin words for the Copper Age, which is good enough for him and many English authors from then on. He offers literary proof that bronze had been in use before iron and copper before bronze. [80]

In 1884 the center of archaeological interest shifted to Italy with the excavation of Remedello and the discovery of the Remedello culture by Gaetano Chierici. According to his 1886 biographers, Luigi Pigorini and Pellegrino Strobel, Chierici devised the term Età Eneo-litica to describe the archaeological context of his findings, which he believed were the remains of Pelasgians, or people that preceded Greek and Latin speakers in the Mediterranean. The age (Età) was:[81]

> "A period of transition from the age of stone to that of bronze (periodo di transizione dall'età della pietra a quella del bronzo)"

Whether intentional or not, the definition was the same as Evans', except that Chierici was adding a term to New Latin. He describes the transition by stating the beginning (litica, or stone age) and the ending (eneo-, or Bronze Age); in English, "the stone-to-bronze period." Shortly after, "Eneolithic" or "Aeneolithic" began turning up in scholarly English as a synonym for "Copper Age." Sir John's own son, Arthur Evans, beginning to come into his own as an archaeologist and already studying Cretan civilization, refers in 1895 to some clay figures of "aeneolithic date" (quotes his).

2.4 Iron age subdivisions

The advent of the Iron Age is marked by the initial use of iron in any region, whether brought in from elsewhere, or by evolution of the smelting process in that region. As the ancient writers considered that they were in the Iron Age, they did not define an end to it. This convention prevailed in modern archaeology as well. Iron is still the major hard material in use in modern civilization. Steel is a vital and indispensable modern industry. Many other suggestions have been made: industrial, machine, plastic, information, etc., but none have been seriously incorporated into the three-age system. Scholarship leaves the question open. Generally in history the Iron Age refers to mainly the 1st millennium BC, no later than the 1st millennium AD. In some cases, however, modern civilization is the Iron Age, as when the African archaeologists hypothesized that Europeans and Middle Easterners brought it to Central and South Africa.

2.5 Dating

The question of the dates of the objects and events discovered through archaeology is the prime concern of any system of thought that seeks to summarize history through the formulation of ages or epochs. An age is defined through comparison of contemporaneous events. Increasingly, the terminology of archaeology is parallel to that of historical method. An event is "undocumented" until it turns up in the archaeological record. Fossils and artifacts are "documents" of the epochs hypothesized. The correction of dating errors is therefore a major concern.

In the case where parallel epochs defined in history were available, elaborate efforts were made to align European and Near Eastern sequences with the datable chronology of Ancient Egypt and other known civilizations. The resulting grand sequence was also spot checked by evidence of calculateable solar or other astronomical events. These methods are only available for the relatively short term of recorded history. Most prehistory does not fall into that category.

Physical science provides at least two general groups of dating methods, stated below. Data collected by these methods is intended to provide an absolute chronology to the framework of periods defined by relative chronology.

2.5.1 Grand systems of layering

The initial comparisons of artifacts defined periods that were local to a site, group of sites or region. Advances made in the fields of seriation, typology, stratification and the associative dating of artifacts and features permitted even greater refinement of the system. The ultimate development is the reconstruction of a global catalogue of layers (or as close to it as possible) with different sections attested in different regions. Ideally once the layer of the artifact or event is known a quick lookup of the layer in the grand system will provide a ready date. This is considered the most reliable method. It is used for calibration of the less reliable chemical methods.

2.5.2 Measurement of chemical change

Any material sample contains elements and compounds that are subject to decay into other elements and compounds. In cases where the rate of decay is predictable and the proportions of initial and end products can be known exactly, consistent dates of the artifact can be calculated. Due to the problem of sample contamination and variability of the natural proportions of the materials in the media, sample analysis in the case where verification can be checked by grand layering systems has often been found to be widely inaccurate. Chemical dates therefore are only considered reliable used in conjunction with other methods. They are collected in groups of data points that form a pattern when graphed. Isolated dates are not considered reliable.

2.6   Other -liths and -lithics

The term Megalithic does not refer to a period of time, but merely describes the use of large stones by ancient peoples from any period. An eolith is a stone that might have been formed by natural process but occurs in contexts that suggest modification by early humans or other primates for percussion.

2.7   Three-age system resumptive table

* Formation of states starts during the Early Bronze Age in Egypt and Mesopotamia and during the Late Bronze Age first empires are founded.

2.8   Criticism

The Three-age System has been criticized since at least the 19th century. Every phase of its development has been contested. Some of the arguments that have been presented against it follow.

2.8.1   Unsound epochalism

In some cases criticism resulted in other, parallel three-age systems, such as the concepts expressed by Lewis Henry Morgan in *Ancient Society*, based on ethnology. These disagreed with the metallic basis of epochization. The critic generally substituted his own definitions of epochs. Vere Gordon Childe said of the early cultural anthropologists:[82]

> "Last century Herbert Spencer, Lewis H. Morgan and Tylor propounded divergent schemes ... they arranged these in a logical order .... They assumed that the logical order was a temporal one.... The competing systems of Morgan and Tylor remained equally unverified—and incompatible—theories."

More recently, many archaeologists have questioned the validity of dividing time into epochs at all. For example, one recent critic, Graham Connah, describes the three-age system as "epochalism" and asserts:[83]

> "So many archaeological writers have used this model for so long that for many readers it has taken on a reality of its own. In spite of the theoretical agonizing of the last half-century, epochalism is still alive and well ... Even in parts of the world where the model is still in common use, it needs to be accepted that, for example, there never was actually such a thing as 'the Bronze Age.'"

2.8.2   Simplisticism

Some view the three-age system as over-simple; that is, it neglects vital detail and forces complex circumstances into a mold they do not fit. Rowlands argues that the division of human societies into epochs based on the presumption of a single set of related changes is not realistic:[84]

> "But as a more rigorous sociological approach has begun to show that changes at the economic, political and ideological levels are not 'all of apiece' we have come to realise that time may be segmented in as many ways as convenient to the researcher concerned."

The three-age system is a relative chronology. The explosion of archaeological data acquired in the 20th century was intended to elucidate the relative chronology in detail. One consequence was the collection of absolute dates. Connah argues:[83]

"As radiocarbon and other forms of absolute dating contributed more detailed and more reliable chronologies, the epochal model ceased to be necessary."

Peter Bogucki of Princeton University summarizes the perspective taken by many modern archaeologists:*[85]

"Although modern archaeologists realize that this tripartite division of prehistoric society is far too simple to reflect the complexity of change and continuity, terms like 'Bronze Age' are still used as a very general way of focusing attention on particular times and places and thus facilitating archaeological discussion."

2.8.3 Eurocentrism

Another common criticism attacks the broader application of the Three-age System as a cross-cultural model for social change. The model was originally designed to explain data from Europe and West Asia, but archaeologists have also attempted to use it to explain social and technological developments in other parts of the world such as the Americas, Australasia, and Africa.*[86] Many archaeologists working in these regions have criticized this application as eurocentric. Graham Connah writes that:*[83]

"... attempts by Eurocentric archaeologists to apply the model to African archaeology have produced little more than confusion, whereas in the Americas or Australasia it has been irrelevant, ..."

Alice B. Kehoe further explains this position as it relates to American archaeology:*[86]

"... Professor Wilson's presentation of prehistoric archaeology*[87] was a European product carried across the Atlantic to promote an American science compatible with its European model."

She goes on to complain of Wilson that "he accepted and reprised the idea that the European course of development was paradigmatic for humankind." *[88] This criticism argues that the different societies of the world underwent social and technological developments in different ways. A sequence of events that describes the developments of one civilization may not necessarily apply to another, in this view. Instead social and technological developments must be described within the context of the society being studied.

2.9 See also

- Atomic Age
- Industrial Age
- Information Age
- List of archaeological periods
- Periodization
- Social Age

2.10 References

[1] Lines 109-201.

[2] Lines 140-155, translator Richmond Lattimore.

[3] Lines 161-169.

[4] Beye, Charles Rowan (Jan 1963). "Lucretius and Progress". *The Classical Journal* **58** (4): 160–169.

[5] De Rerum Natura, Book V, about Line 800 ff. The translator is Ronald Latham.

[6] *De Rerum Natura*, Book V, around Line 1200 ff.

[7] *De Rerum Natura*, Book V around Line 940 ff.

[8] Goodrum 2008, p. 483

[9] Goodrum 2008, p. 494

[10] Goodrum 2008, p. 495

[11] Goodrum 2008, p. 496.

[12] Hamy 1906, pp. 249–251

[13] Hamy 1906, p. 246

[14] Hamy 1906, p. 252

[15] Hamy 1906, p. 259: "c'est a Michel Mercatus, Médecin de Clément VIII, que la première idée est duë..."

[16] Rowley-Conwy 2007, p. 40

[17] Rowley-Conwy 2007, p. 22

[18] Rowley-Conwy 2007, p. 36

[19] Rowley-Conwy 2007, Front Matter, Abbreviations

[20] Malina & Vašíček 1990, p. 37

[21] Rowley-Conwy 2007, p. 38

[22] Gräslund 1987, p. 23

[23] Gräslund 1987, pp. 22, 28

[24] Gräslund 1987, pp. 18–19

[25] Rowley-Conwy 2007, pp. 298–301

[26] Gräslund 1987, p. 24

[27] Thomsen, Christian Jürgensen (1836). "Kortfattet udsigt over midesmaeker og oldsager fra Nordens oldtid". In Rafn, C.C. *Ledetraad til Nordisk Oldkyndighed* (in Danish). Copenhagen: Kongelige Nordiske Oldskriftselskab.

[28] This was not the museum guidebook, which was written by Julius Sorterup, an assistant of Thomsen, and published in 1846. Note that translations of Danish organizations and publications tend to vary somewhat.

[29] Lubbock 1865, pp. 2–3

[30] Lubbock 1865, pp. 336–337

[31] Lubbock 1865, p. 472

[32] "Reviews". *The Medical Times and Gazette: A Journal of Medical Science, Literature, Criticism and News* (London: John Churchill and Sons) **II**. Aug 6, 1870.

[33] Westropp 1866, p. 288

[34] Westropp 1866, p. 291

[35] Westropp 1866, p. 290

[36] Westropp 1872, p. 41

[37] Westropp 1872, p. 45

[38] Westropp 1872, p. 53

[39] Evans 1872, p. 12

[40] Taylor, Isaac (1889). *The Origin of the Aryans. An Account of the Prehistoric Ethnology and Civilisation of Europe.* New York: C. Scribner's sones. p. 60.

[41] Haeckel, Ernst Heinrich Philipp August; Lankester, Edwin Ray (1876). *The history of creation, or, The development of the earth and its inhabitants by the action of natural causes : a popular exposition of the doctrine of evolution in general, and of that of Darwin, Goethe, and Lamarck in particular.* New York: D. Appleton. p. 15.

[42] Brown 1893, p. 66

[43] Piette 1895, p. 236: "Entre le paléolithique et le neolithique, il y a une large et profonde lacune, un grand hiatus; ..."

[44] Piette 1895, p. 237

[45] Piette 1895, p. 239: "J'ai eu la bonne fortune découvrir les restes de cette époque ignorée qui sépara l'àge magdalénien de celui des haches en pierre polie ... ce fut, au Mas-d'Azil, en 1887 et en 1888 que je fis cette découverte."

[46] Brown 1893, pp. 74–75.

[47] Stjerna 1910, p. 2

[48] Stjerna 1910, p. 10

[49] Stjerna 1910, p. 12: "... a persisté pendant la période paléolithique récente et même pendant la période protonéolithique."

[50] Stjerna 1910, p. 12

[51] Obermaier, Hugo (1924). *Fossil man in Spain.* New Haven: Yale University Press. p. 322.

[52] Farrand, W.R. (1990). "Origins of Quaternary-Pleistocene-Holocene Stratigraphic Terminology". In Laporte, Léo F. *Establishment of a Geologic Framework for Paleoanthropology.* Special Paper 242. Boulder: Geological Society of America. pp. 16–18

[53] Dawkins 1880, p. 3

[54] Dawkins 1880, p. 124

[55] Dawkins 1880, p. 247

[56] Dawkins 1880, p. 183

[57] Dawkins 1880, p. 181

[58] Dawkins 1880, p. 178

[59] Geikie, James (1881). *Prehistoric Europe: A Geological Sketch.* London: Edward Stanford..

[60] Sollas, William Johnson (1911). *Ancient hunters: and their modern representatives.* London: Macmillan and Co. p. 130.

[61] "On an Earlier and Later Period in the Stone Age". *The Gentleman's Magazine*: 548. May 1862.

[62] Wallace, Alfred Russel (1864). "The Origin of Human Races and the Antiquity of Man Deduced From the Theory of "Natural Selection"". *Journal of the Anthropological Society of London* **2**.

[63] Lubbock 1865, p. 481

[64] Howarth, H.H. (1875). "Report on the Stockholm Meeting of the International Congress of Anthropology and Prehistoric Archaeology". *Journal of the Royal Anthropological Institute of Great Britain and Ireland* **IV**: 347.

[65] Chambers, William and Robert (December 20, 1879). "Pre-historic Records". *Chambers's Journal* **56** (834): 805–808.

[66] Garašanin, M. (1925). "Cambridge Ancient Histiry" |contribution= ignored (help)

[67] Childe 1951, p. 44

[68] Childe 1951, pp. 34–35

[69] Childe 1951, p. 14

[70] Childe 1951, p. 42

[71] Childe, who was writing for the masses, did not make use of critical apparatus and offered no attributions in his texts. This practice led to the erroneous attribution of the entire three-age system to him. Very little of it originated with him. His synthesis and expansion of its detail is however attributable to his presentations.

[72] Lubbock 1865, p. 323

[73] Dawkins, W. Boyd (July 1866). "On the Habits and Conditions of the Two earliest known Races of Men". *Quarterly Journal of Science* **3**: 344.

[74] "Kenyon Institute". Retrieved 31 May 2011.

[75] Howorth, H.H. (1875). "Report of the Stockholm Meeting of the International Congress of Anthropology and Prehistoric Archaeology". *Journal of the Anthropological Institute of Great Britain and Ireland* (London: AIGBI) **IV**: 354–355.

[76] Evans 1881, p. 456

[77] Evans 1881, p. 410

[78] Evans 1881, p. 474

[79] Evans 1881, p. 2

[80] Evans 1881, Chapter 1

[81] Pigorini, Luigi; Strobel, Pellegrino (1886). *Gaetano Chierici e la paletnologia italiana* (in Italian). Parma: Luigi Battei. p. 84.

[82] Childe, V. Gordon; Patterson, Thomas Carl; Orser, Charles E. (2004). *Foundations of social archaeology: selected writings of V. Gordon Childe*. Walnut Creek, California: AltaMira Press. p. 173.

[83] Connah 2010, pp. 62–63

[84] Kristiansen & Rowlands 1998, p. 47

[85] Bogucki 2008

[86] Browman & Williams 2002, p. 146

[87] A predecessor of Lubbock working from the original Danish conception of the three ages.

[88] Browman & Williams 2002, p. 147

2.11 Bibliography

- Browman, David L.; Williams, Steven (2002). *New Perspectives on the Origins of Americanist Archaeology*. Tuscaloosa: University of Alabama Press.

- Brown, J. Allen (1893). *The Journal of the Anthropological Institute of Great Britain and Ireland* **XXII**: 66–98. Missing or empty |title= (help); |contribution= ignored (help)

- Bogucki, Peter (2008). "Northern and Western Europe: Bronze Age". *Encyclopedia of Archaeology*. New York: Academic Press. pp. 1216–1226.

- Childe, V. Gordon (1951). *Man Makes Himself* (3rd ed.). Mentor Books (New American Library of World Literature, Inc.).

- Connah, Graham (2010). *Writing About Archaeology*. Cambridge University Press.

- Dawkins, William Boyd (1880). *The Three Pleistocene Strata: Early Man in Britain and his place in the Tertiary Period*. London: MacMillan and Co.

- Evans, John (1872). *The ancient stone implements, weapons and ornaments, of Great Britain*. New York: D. Appleton and Company.

- Evans, John (1881). *The Ancient Bronze Implements, Weapons, and Ornaments of Great Britain and Ireland*. London: Longmans Green & Co.

- Goodrum, Matthew R. (2008). "Questioning Thunderstones and Arrowheads: The Problem of Recognizing and Interpreting Stone Artifacts in the Seventeenth Century". *Early Science and Medicine* **13** (5): 482–508. doi:10.1163/157338208X345759.

- Gräslund, Bo (1987). *The Birth of Prehistoric Chronology. Dating methods and dating systems in nineteenth-century Scandinavian archeology*. Cambridge: Cambridge University Press.

- Hamy, M.E.T. (1906). "Matériaux pour servir à l'histoire de l'archéologie préhistorique". *Revue archéologique*. 4th Series (in French) **7** (March–April): 239–259.

- Heizer, Robert F. (1962). "The background of Thomsen's Three-Age System". *Technology and Culture* **3** (3): 259–266. doi:10.2307/3100819.

- Kristiansen, Kristian; Rowlands, Michael (1998). *Social Transformations in Archaeology: global and local persepectives*. London: Routledge.

- Lubbock, John (1865). *Pre-historic times. as illustrated by ancient remains, and the manners and customs of modern savages*. London & Edinburgh: Williams and Norgate.

- Malina, Joroslav; Vašíček, Zdenek (1990). *Archaeology yesterday & today: The development of archaeology in the sciences & humanities*. Cambridge: Cambridge University Press.

- Piette, Edouard (1895). "Hiatus et lacune. Vestiges de la période de transition dans la grotte du Mas-d'Azil". *Bulletin de la Societé d'anthropologie de Paris* (in French) **6** (6): 235–267. doi:10.3406/bmsap.1895.5585. |contribution= ignored (help)

- Rowley-Conwy, Peter (2007). *From Genesis to Prehistory: The Archaeological Three Age System and its Contested Reception in Denmark, Britain, and Ireland*. Oxford Studies in the History of Archaeology. Oxford, New York: Oxford University Press.

- Rowley-Conwy, Peter (2006). "The Concept of Prehistory and the Invention of the Terms 'Prehistoric' and 'Prehistorian': the Scandinavian Origin, 1833 — 1850". *European Journal of Archaeology* **9** (1): 103–130. doi:10.1177/1461957107077709.

- Sterjna, Knut (1910). *L'Anthropologie* (in French) (Paris) **XXI**: 1–34. Missing or empty |title= (help); |contribution= ignored (help)

- Trigger, Bruce (2006). *A History of Archaeological thought* (2nd ed.). Oxford: Cambridge University Press.

- Westropp, Hodder M. (1866). "Memoirs Read Before the Society". Publications of the Anthropological Society of London **II**. London: Anthropological Society of London. pp. 288–294 |contribution= ignored (help)

- Westropp, Hodder M. (1872). *Pre-Historic Phases; or, Introductory Essays on Pre-Historic Archaeology*. London: Bell & Daldy.

Hesiod inspired by the Muse, Gustave Moreau, 1891

(I)

T. LUCRETII CARI
DE
RERUM NATURA
Liber Primus.

ENEADUM genetrix, hominum di-
vumque voluptas,
Alma Venus, coeli subter labentia
signa
Quæ mare navigerum, quæ terras fru-
giferenteis.
Concelebras : per te quoniam genus omne animantum
Concipitur, vifitque exortum lumina folis :
Te, Dea, te fugiunt venti, te nubila coeli,
Adventumque tuum : tibi fuaveis dædala tellus
Summittit flores, tibi rident æquora ponti,
Placatumque nitet diffufo lumine coelum.
Nam fimul ac fpecies patefacta 'ft verna diei,
Et referata viget genitabilis aura FavonI :
Aeriæ primum volucres te, diva, tuumque
Significant initum percuffæ corda tua vi.
Inde feræ pecudes perfultant pabula læta,
Et rapidos tranant amneis : ita capta lepore,
Illecebrifque tuis omnis natura animantum

B   Te

Page 1 Chapter 1 of De Rerum Natura, 1675, dedicating the poem to Alma Venus

Michele Mercati, Commemorative Medal.

Thomsen explaining the Three-age System to visitors at the Museum of Northern Antiquities, then at the Christiansborg Palace, in Copen-hagen, 1846. Drawing by Magnus Petersen, Thomsen's illustrator. [16]

Reconstructed Iron Age home in Spain

Bone harpoon studded with microliths, a Mode 5 composite hunting implement.

Mas-d'Azil Grotto

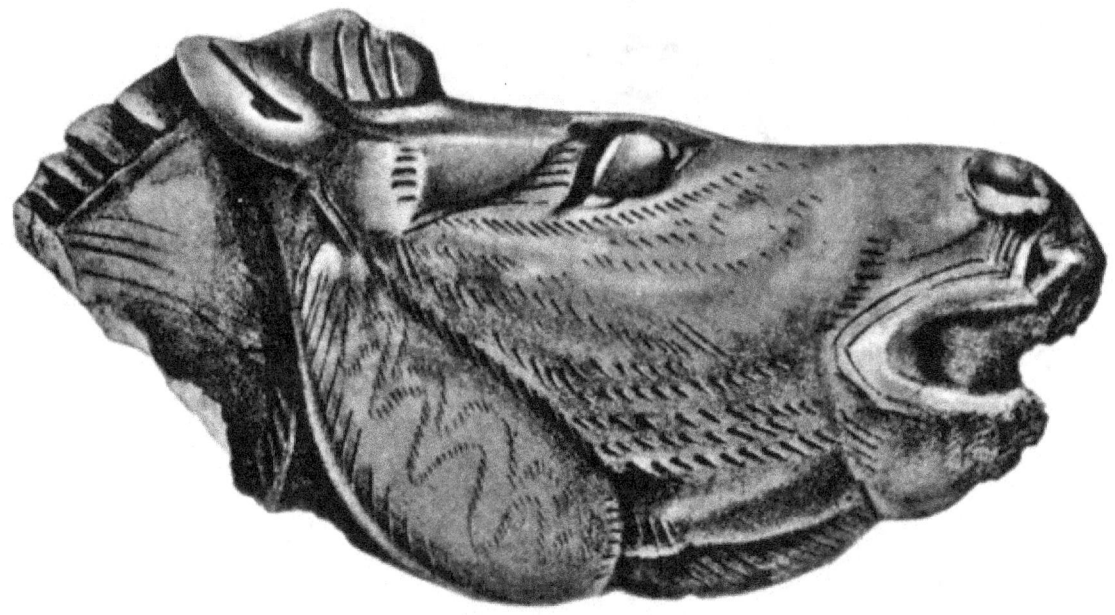

Small Magdalenian carving representing a horse.

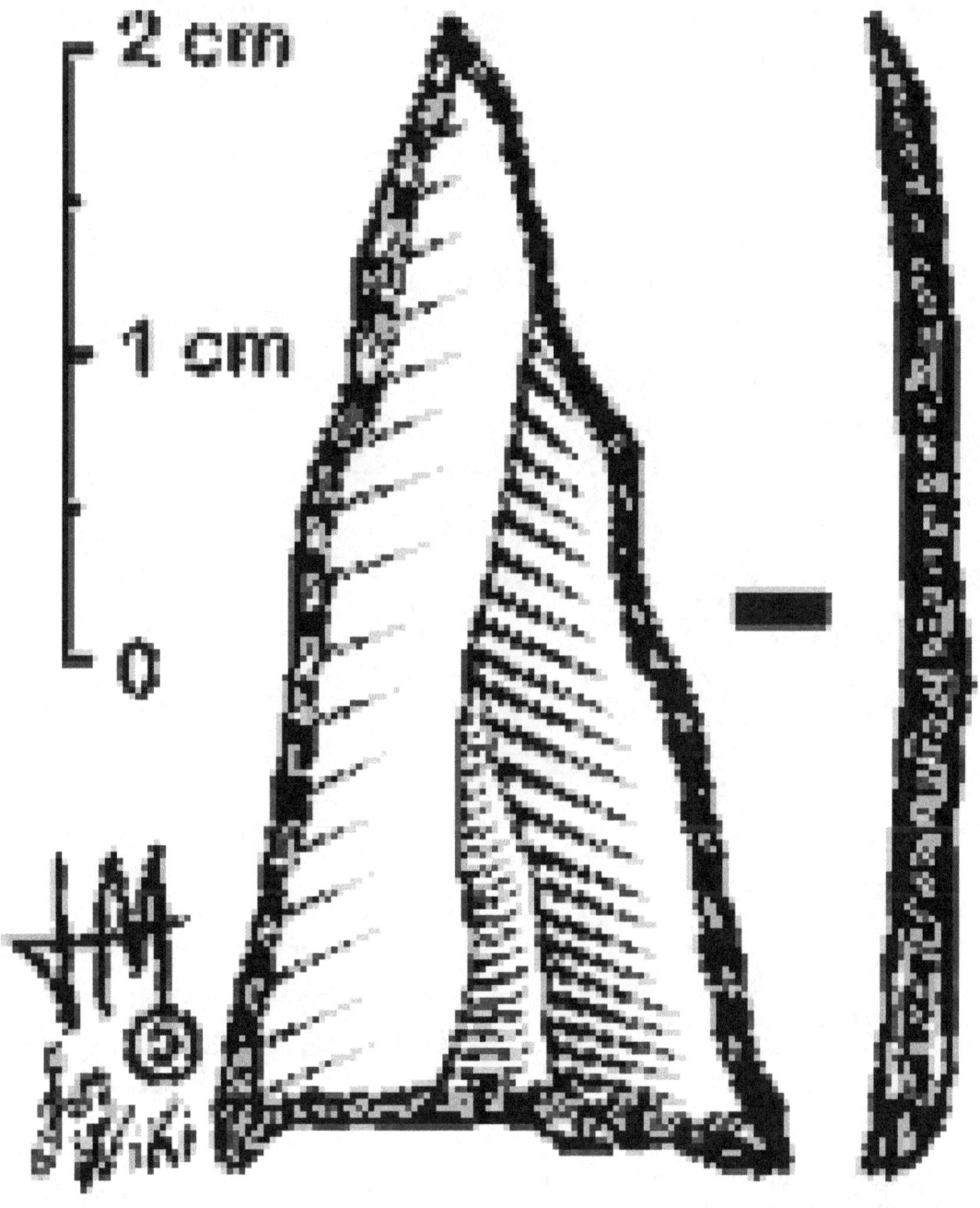

Tardenoisian Mode 5 point—Mesolithic or Epipaleolithic?

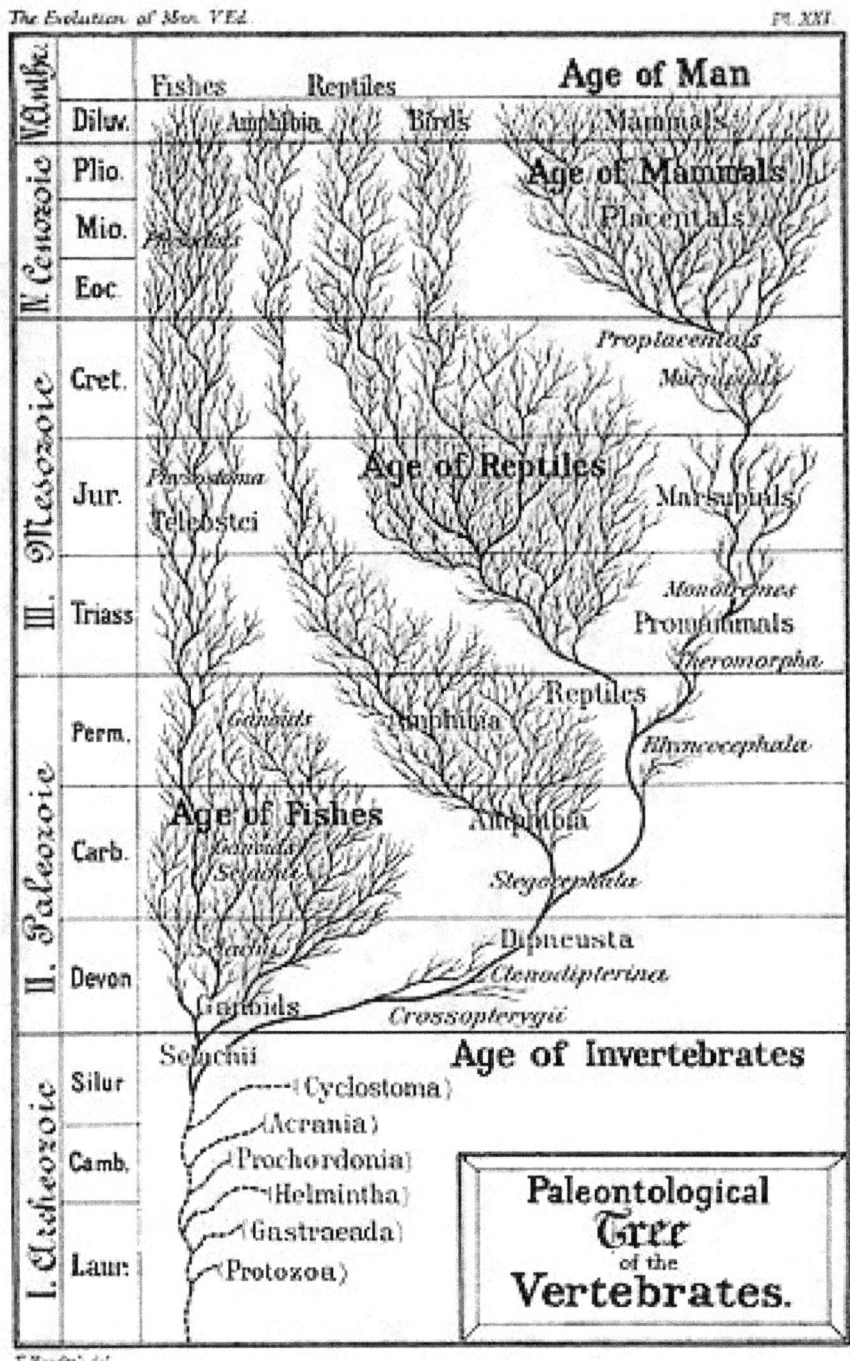

Haeckel's tree growing through the layers. In geology, the tripartite division did not stand the test of time.

Chapter 3

Ancient Near East

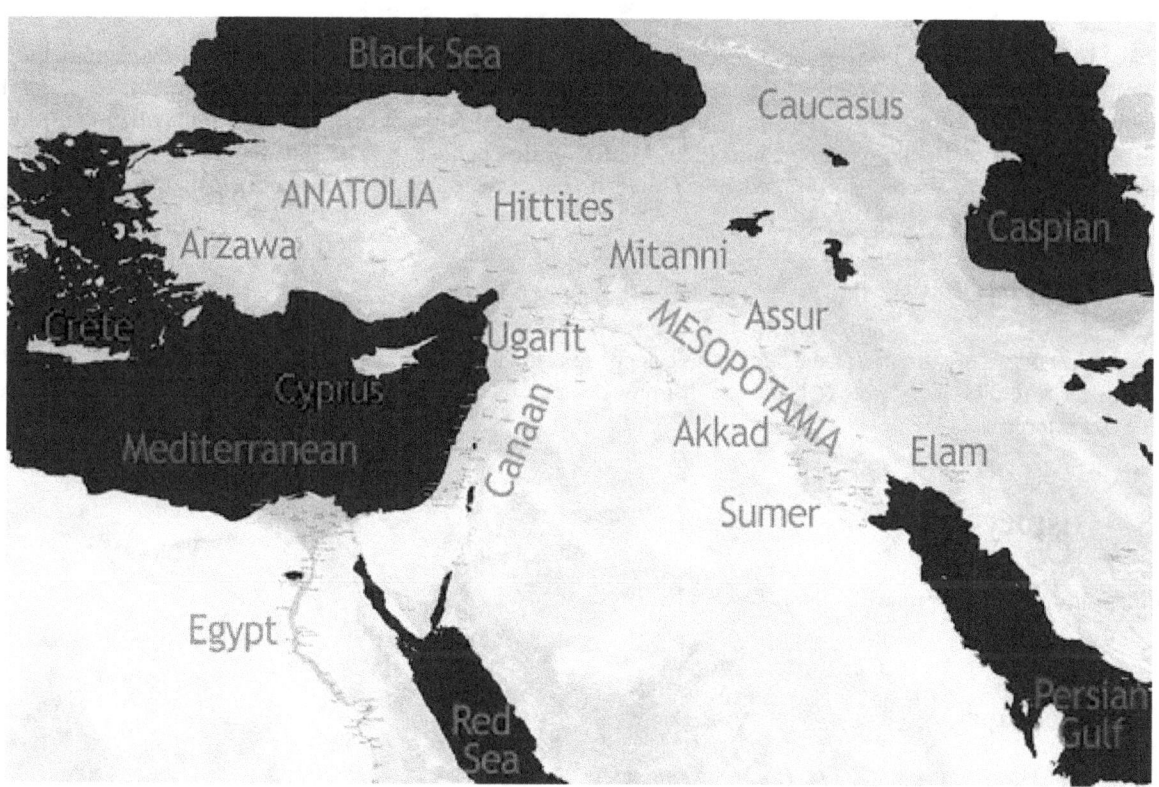

Overview map of the ancient Near East

The **ancient Near East** was the home of early civilizations within a region roughly corresponding to the modern Middle East: Mesopotamia (modern Iraq, southeast Turkey, southwest Iran, northeastern Syria and Kuwait),*[1] ancient Egypt, ancient Iran (Elam, Media, Parthia and Persia), Anatolia/Asia Minor and Armenian Highlands (Turkey's Eastern Anatolia Region, Armenia, northwestern Iran, southern Georgia, and western Azerbaijan),*[2] the Levant (modern Syria, Lebanon, Palestine, Israel, and Jordan), Cyprus and the Arabian Peninsula. The ancient Near East is studied in the fields of Near Eastern archaeology and ancient history. It begins with the rise of Sumer in the 4th millennium BC, though the date it ends varies: either covering the Bronze Age and the Iron Age in the region, until the conquest by the Achaemenid Empire in the 6th century BC or Alexander the Great in the 4th century BC.

The ancient Near East is considered the cradle of civilization. It was here that intensive year-round agriculture was first practiced, leading to the rise of the first dense urban settlements and the development of many familiar institutions of civilization, such as social stratification, centralized government and empires, organized religion and organized warfare. It also saw the creation of the first writing system and law codes, early advances that laid the foundations of astronomy and mathematics, and the invention of the wheel.

3.1 The concept of Near East

Main article: Near East

The term "ancient Near East" utilizes the 19th-century distinction between Near East and Far East as global regions of interest to the British Empire. The distinction began during the Crimean War. The last major exclusive partition of the east between these two terms was current in diplomacy in the late 19th century with the Hamidian Massacres of the Armenians and Assyrians by the Ottoman Empire in 1894-1896 and the Sino-Japanese War of 1894-1895. The two theatres were described by the statesmen and advisors of the British Empire as "the Near East" and "the Far East." Shortly they were to share the stage with Middle East, which came to prevail in the 20th century and continues in modern times.

As Near East had meant the lands of the Ottoman Empire at roughly maximum extent, on the fall of that empire the use of Near East in diplomacy was reduced significantly in favor of Middle East. In the meanwhile, ancient Near East had become distinct. The Near East ruled by the Ottoman Empire ranged from Vienna to the north to the tip of the Arabian Peninsula to the south, from Egypt in the west to the borders of Iraq in the east. The 19th-century archaeologists added Iran to their definition, which was never under the Ottomans, but excluded all of Europe, and generally Egypt.

3.2 Periodization

Ancient Near East periodization is the attempt to categorize or divide time into discrete named blocks, or eras, of the Near east. The result is a descriptive abstraction that provides a useful handle on Near East periods of time with relatively stable characteristics.

3.3 History

Further information: Timeline of Middle Eastern History

3.3.1 Prehistory

Main article: ASPRO chronology

- Paleolithic
- Epipaleolithic and mesolithic
  - Kebaran culture
  - Natufian culture
- Pre-pottery Neolithic A
- Pre-pottery Neolithic B
- Pre-pottery Neolithic C
- Pottery Neolithic

3.3.2 Chalcolithic

Early Mesopotamia

The Uruk period (c. 4000 to 3100 BC) existed from the protohistoric Chalcolithic to Early Bronze Age period in the history of Mesopotamia, following the Ubaid period.*[3] Named after the Sumerian city of Uruk, this period saw the emergence of urban life in Mesopotamia. It was followed by the Sumerian civilization.*[4] The late Uruk period (34th to 32nd centuries) saw the gradual emergence of the cuneiform script and corresponds to the Early Bronze Age.

3.3.3 Bronze Age

Further information: Short chronology timeline

Early Bronze Age

Sumer & Akkad   Sumer, located in southern Mesopotamia, is the earliest known civilization in the world. It lasted from the first settlement of Eridu in the Ubaid period (late 6th millennium BC) through the Uruk period (4th millennium BC) and the Dynastic periods (3rd millennium BC) until the rise of Assyria and Babylon in the late 3rd millennium BC and early 2nd millennium BC respectively. The Akkadian Empire, founded by Sargon the Great, lasted from the 24th to the 21st century BC, and was regarded by many as the world's first Empire. The Akkadians eventually fragmented into Assyria and Babylonia.

Elam   Ancient Elam lay to the east of Sumer and Akkad, in the far west and southwest of modern-day Iran, stretching from the lowlands of Khuzestan and Ilam Province. In the Old Elamite period c. 3200 BC, it consisted of kingdoms on the Iranian plateau, centered in Anshan, and from the mid-2nd millennium BC, it was centered in Susa in the Khuzestan lowlands. Elam was absorbed into the Assyrian Empire in the 9th to 7th centuries BC, however the civilization endured up until 539 BC when it was finally overrun by the Iranian Persians. The Proto-Elamite civilization existed during the time of c. 3200 BC to 2700 BC when Susa, the later capital of the Elamites began to receive influence from the cultures of the Iranian plateau. In archaeological terms this corresponds to the late Banesh period. This civilization is recognized as the oldest in Iran and was largely contemporary with its neighbour, Sumerian civilization. The Proto-Elamite script is an Early Bronze Age writing system briefly in use for the ancient Elamite language (which was a Language isolate) before the introduction of Elamite Cuneiform.

The Amorites   The Amorites were a nomadic Semitic people who occupied the country west of the Euphrates from the second half of the third millennium BC. In the earliest Sumerian sources, beginning about 2400 BC, the land of the Amorites ("the *Mar.tu* land") is associated with the West, including Syria and Canaan, although their ultimate origin may have been Arabia.*[5] They ultimately settled in Mesopotamia, ruling Isin, Larsa, and later Babylon.

Middle Bronze Age

- Assyria after enduring a short period of Mitanni domination, emerged as a great power from the assecion of Ashur-uballit I in 1365 BC to the death of Tiglath-Pileser I in 1076 BC. Assyria rivalled Egypt during this period, and dominated much of the near east.

- Babylonia, originally founded as a state by Amorite tribes, found itself under the rule of Kassites for 435 years. The nation stagnated during the Kassite period, and Babylonia often found itself under Assyrian or Elamite domination.

- Canaan: Ugarit, Kadesh, Megiddo, Kingdom of Israel

- The Hittite Empire was founded some time after 2000 BC, and existed as a major power, dominating Asia Minor and the Levant until 1200 BC when it was first overrun by the Phrygians, and then appropriated by Assyria.

Late Bronze Age

The Hurrians lived in northern Mesopotamia and areas to the immediate east and west, beginning approximately 2500 BC. They probably originated in the Caucasus and entered from the north, but this is not certain. Their known homeland was centred in Subartu, the Khabur River valley, and later they established themselves as rulers of small kingdoms throughout northern Mesopotamia and Syria. The largest and most influential Hurrian nation was the kingdom of Mitanni. The Hurrians played a substantial part in the History of the Hittites.

Ishuwa was an ancient kingdom in Anatolia. The name is first attested in the second millennium BC, and is also spelled Išuwa. In the classical period the land was a part of Armenia. Ishuwa was one of the places where agriculture developed very early in the Neolithic. Urban centres emerged in the upper Euphrates river valley around 3500 BC. The first states followed in the third millennium BC. The name Ishuwa is not known until the literate period of the second millennium BC. Few literate sources from within Ishuwa have been discovered and the primary source material comes from Hittite texts. To the west of Ishuwa lay the kingdom of the Hittites, and this nation was an untrustworthy neighbour. The Hittite king Hattusili I (c. 1600 BC) is reported to have marched his army across the Euphrates river and destroyed the cities there. This corresponds well with burnt destruction layers discovered by archaeologists at town sites in Ishuwa of roughly the same date. After the end of the Hittite empire in the early 12th century BC a new state emerged in Ishuwa. The city of Malatya became the centre of one of the so-called Neo-Hittite kingdom. The movement of nomadic people may have weakened the kingdom of Malatya before the final Assyrian invasion. The decline of the settlements and culture in Ishuwa from the 7th century BC until the Roman period was probably caused by this movement of people. The Armenians later settled in the area since they were natives of the Armenian Plateau and related to the earlier inhabitants of Ishuwa.

Kizzuwatna is the name of an ancient kingdom of the second millennium BC. It was situated in the highlands of south-eastern Anatolia, near the Gulf of İskenderun in modern-day Turkey. It encircled the Taurus Mountains and the Ceyhan river. The centre of the kingdom was the city of Kummanni, situated in the highlands. In a later era, the same region was known as Cilicia.

Luwian is an extinct language of the Anatolian branch of the Indo-European language family. Luwian speakers gradually spread through Anatolia and became a contributing factor to the downfall, after c. 1180 BC, of the Hittite Empire, where it was already widely spoken. Luwian was also the language spoken in the Neo-Hittite states of Syria, such as Melid and Carchemish, as well as in the central Anatolian kingdom of Tabal that flourished around 900 BC. Luwian has been preserved in two forms, named after the writing systems used to represent them: Cuneiform Luwian, and Hieroglyphic Luwian.

Mari was an ancient Sumerian and Amorite city, located 11 kilometres north-west of the modern town of Abu Kamal on the western bank of Euphrates river, some 120 km southeast of Deir ez-Zor, Syria. It is thought to have been inhabited since the 5th millennium BC, although it flourished from 2900 BC until 1759 BC, when it was sacked by Hammurabi.

Mitanni was a Hurrian kingdom in northern Mesopotamia from c. 1500 BC, at the height of its power, during the 14th century BC, encompassing what is today southeastern Turkey, northern Syria and northern Iraq (roughly corresponding to Kurdistan), centred on the capital Washukanni whose precise location has not yet been determined by archaeologists. The Mitanni kingdom is thought to have been a feudal state led by a warrior nobility of Indo-Aryan descent, who invaded the Levant region at some point during the 17th century BC, their influence apparent in a linguistic superstratum in Mitanni records. The spread to Syria of a distinct pottery type associated with the Kura-Araxes culture has been connected with this movement, although its date is somewhat too early.*[6] Yamhad was an ancient Amorite kingdom. A substantial Hurrian population also settled in the kingdom, and the Hurrian culture influenced the area. The kingdom was powerful during the Middle Bronze Age, c. 1800-1600 BC. Its biggest rival was Qatna further south. Yamhad was finally destroyed by the Hittites in the 16th century BC.

The Aramaeans were a Semitic (West Semitic language group), semi-nomadic and pastoralist people who had lived in upper Mesopotamia and Syria. Aramaeans have never had a unified empire; they were divided into independent kingdoms all across the Near East. Yet to these Aramaeans befell the privilege of imposing their language and culture upon the entire Near East and beyond, fostered in part by the mass relocations enacted by successive empires, including the Assyrians and Babylonians. Scholars even have used the term 'Aramaization' for the Assyro-Babylonian peoples' languages and cultures, that have become Aramaic-speaking.*[7]

The Sea peoples is the term used for a confederacy of seafaring raiders of the second millennium BC who sailed into the eastern shores of the Mediterranean, caused political unrest, and attempted to enter or control Egyptian territory during the late 19th dynasty, and especially during Year 8 of Ramesses III of the 20th Dynasty.*[8] The Egyptian Pharaoh Merneptah explicitly refers to them by the term "the foreign-countries (or 'peoples'*[9]) of the sea" *[10]*[11] in his Great Karnak Inscription.*[12] Although some scholars believe that they "invaded" Cyprus, Hatti and the Levant, this hypothesis is disputed.*[13]

Bronze Age collapse The *Bronze Age collapse* is the name given by those historians who see the transition from the Late Bronze Age to the Early Iron Age as violent, sudden and culturally disruptive, expressed by the collapse of palace economies of the Aegean and Anatolia, which were replaced after a hiatus by the isolated village cultures of the Dark Age period in history of the ancient Middle East. Some have gone so far as to call the catalyst that ended the Bronze Age a "catastrophe".*[14] The Bronze Age collapse may be seen in the context of a technological history that saw the slow, comparatively continuous spread of iron-working technology in the region, beginning with precocious iron-working in what is now Romania in the 13th and 12th centuries.*[15] The cultural collapse of the Mycenaean kingdoms, the Hittite Empire in Anatolia and Syria, and the Egyptian Empire in Syria and Israel, the scission of long-distance trade contacts and sudden eclipse of literacy occurred between 1206 and 1150 BC. In the first phase of this period, almost every city between Troy and Gaza was violently destroyed, and often left unoccupied thereafter (for example, Hattusas, Mycenae, Ugarit). The gradual end of the Dark Age that ensued saw the rise of settled Neo-Hittite and Aramaean kingdoms of the mid-10th century BC, and the rise of the Neo-Assyrian Empire.

3.3.4 Iron Age

During the Early Iron Age, from 911 BC, the Neo-Assyrian Empire arose, vying with Babylonia and other lesser powers for dominance of the region, though not until the reforms of Tiglath-Pileser III in the 8th century BC,*[16]*[17] did it become a powerful and vast empire. In the Middle Assyrian period of the Late Bronze Age, Assyria had been a kingdom of northern Mesopotamia (modern-day northern Iraq), competing for dominance with its southern Mesopotamian rival Babylonia. From 1365-1076 it had been a major imperial power, rivalling Egypt and the Hittite Empire. Beginning with the campaign of Adad-nirari II, it became a vast empire, overthrowing 25th dynasty Egypt and conquering Egypt, the Middle East, and large swathes of Asia Minor, ancient Iran, the Caucasus and east Mediterranean. The Neo-Assyrian Empire succeeded the Middle Assyrian period (14th to 10th century BC). Some scholars, such as Richard Nelson Frye, regard the Neo-Assyrian Empire to be the first real empire in human history.*[18] During this period, Aramaic was also made an official language of the empire, alongside the Akkadian language.*[18]

The states of the Neo-Hittite kingdoms were Luwian, Aramaic and Phoenician-speaking political entities of Iron Age northern Syria and southern Anatolia that arose following the collapse of the Hittite Empire around 1180 BC and lasted until roughly 700 BC. The term "Neo-Hittite" is sometimes reserved specifically for the Luwian-speaking principalities like Melid (Malatya) and Karkamish (Carchemish), although in a wider sense the broader cultural term "Syro-Hittite" is now applied to all the entities that arose in south-central Anatolia following the Hittite collapse – such as Tabal and Quwê – as well as those of northern and coastal Syria.*[19]

Urartu was an ancient kingdom of Armenia and North Mesopotamia*[20] which existed from c. 860 BC, emerging from the Late Bronze Age until 585 BC. The Kingdom of Urartu was located in the mountainous plateau between Asia Minor, the Iranian Plateau, Mesopotamia, and the Caucasus mountains, later known as the Armenian Highland, and it centered on Lake Van (present-day eastern Turkey). The name corresponds to the Biblical *Ararat*.

The term Neo-Babylonian Empire refers to Babylonia under the rule of the 11th ("Chaldean") dynasty, from the revolt of Nabopolassar in 623 BC until the invasion of Cyrus the Great in 539 BC (Although the last ruler of Babylonia (Nabonidus) was in fact from the Assyrian city of Harran and not Chaldean), notably including the reign of Nebuchadrezzar II. Through the centuries of Assyrian domination, Babylonia enjoyed a prominent status, and revolted at the slightest indication that it did not. However, the Assyrians always managed to restore Babylonian loyalty, whether through granting of increased privileges, or militarily. That finally changed in 627 BC with the death of the last strong Assyrian ruler, Ashurbanipal, and Babylonia rebelled under Nabopolassar the Chaldean a few years later. In alliance with the Medes and Scythians, Nineveh was sacked in 612 and Harran in 608 BC, and the seat of empire was again transferred to Babylonia. Subsequently, the Medes controlled much of the Ancient Near East from their base in Ecbatana (modern-day Hamadan, Iran), most notably most of what is now Turkey, Iran, Iraq, and the South Caucasus.

Following the fall of the Medes, the Achaemenid Empire was the first of the Persian Empires to rule over most of the Near East and far beyond, and the second great Iranian empire (after the Median Empire). At the height of its power, encompassing approximately 7.5 million square kilometers, the Achaemenid Empire was territorially the largest empire of classical antiquity, and the first world empire. It spanned three continents (Europe, Asia, and Africa), including apart from its core in modern-day Iran, the territories of modern Iraq, the Caucasus (Armenia, Georgia, Azerbaijan, Dagestan, Abkhazia), Asia Minor (Turkey), Thrace, Bulgaria, Greece, many of the Black Sea coastal regions, northern Saudi Arabia, Jordan, Israel, Lebanon, Syria, Afghanistan, Central Asia, parts of Pakistan, and all significant population centers of ancient Egypt as far west as Libya.[21] It is noted in western history as the foe of the Greek city states in the Greco-Persian Wars, for freeing the Israelites from their Babylonian captivity, and for instituting Aramaic as the empire's official language.

3.4   Religions

Main article: Religions of the ancient Near East

Ancient civilizations in the Near East were deeply influenced by their spiritual beliefs, which generally did not distinguish between heaven and Earth.[22] They believed that divine action influenced all mundane matters, and also believed in divination (ability to predict the future).[22] Omens were often inscribed in ancient Egypt and Mesopotamia, as were records of major events.[22]

3.5   See also

- Ancient Near East studies

- Ancient history

- Cities of the ancient Near East

- History of pottery in the Southern Levant

3.6   References

[1] "Daily Life In Ancient Mesopotamia" . Retrieved 28 February 2015.

[2] "Armenian Highland" . Retrieved 28 February 2015.

[3] Sumer and the Sumerians, by Harriet E. W. Crawford, p 69

[4] Sumer and the Sumerians, by Harriet E. W. Crawford, p 75

[5] Amorite *Encyclopædia Britannica*

[6] James P. Mallory, "Kuro-Araxes Culture" , *Encyclopedia of Indo-European Culture*, Fitzroy Dearborn, 1997.

[7] See page 9.

[8] A convenient table of sea peoples in hieroglyphics, transliteration and English is given in the dissertation of Woodhuizen, 2006, who developed it from works of Kitchen cited there.

[9] As noted by Gardiner V.1 p.196, other texts have ḫ3ty.w "foreign-peoples"; both terms can refer to the concept of "foreigners" as well. Zangger in the external link below expresses a commonly held view that "sea peoples" does not translate this and other expressions but is an academic innovation. The Woudhuizen dissertation and the Morris paper identify Gaston Maspero as the first to use the term "peuples de la mer" in 1881.

[10] Gardiner V.1 p.196.

[11] Manassa p.55.

[12] Line 52. The inscription is shown in Manassa p.55 plate 12.

[13] Several articles in Oren.

[14] Drews, Robert (1995). *The End of the Bronze Age: Changes in Warfare and the Catastrophe CA 1200 B.C.* United States: Princeton University Press. p. 264. ISBN 978-0-691-02591-9.

[15] See A. Stoia and the other essays in M.L. Stig Sørensen and R. Thomas, eds., *The Bronze Age—Iron Age Transition in Europe* (Oxford) 1989, and T.H. Wertime and J.D. Muhly, *The Coming of the Age of Iron* (New Haven) 1980.

[16] Assyrian Eponym List

[17] Tadmor, H. (1994). *The Inscriptions of Tiglath-Pileser III, King of Assyria.*pp.29

[18] Frye, Richard N. (1992). "Assyria and Syria: Synonyms". *PhD., Harvard University.* Journal of Near Eastern Studies. And the ancient Assyrian empire, was the first real, empire in history. What do I mean, it had many different peoples included in the empire, all speaking Aramaic, and becoming what may be called, "Assyrian citizens." That was the first time in history, that we have this. For example, Elamite musicians, were brought to Nineveh, and they were 'made Assyrians' which means, that Assyria, was more than a small country, it was the empire, the whole Fertile Crescent.

[19] Hawkins, John David; 1982a. "Neo-Hittite States in Syria and Anatolia" in *Cambridge Ancient History* (2nd ed.) 3.1: 372-441. Also: Hawkins, John David; 1995. "The Political Geography of North Syria and South-East Anatolia in the Neo-Assyrian Period" in *Neo-Assyrian Geography*, Mario Liverani (ed.), Università di Roma "La Sapienza", Dipartimento di Scienze storiche, archeologiche e anthropologiche dell' Antichità, Quaderni di Geografia Storica 5: Roma: Sargon srl, 87-101.

[20] *Urartu* article, Columbia Electronic Encyclopedia, 2007

[21] Full translation of the Behistun Inscription

[22] Lamberg-Karlovsky, C. C. and Jeremy A. Sabloff (1979). *Ancient Civilizations: The Near East and Mesoamerica.* Benjamin/Cummings Publishing. p. 4.

3.7 Further reading

- Fletcher, Banister; Cruickshank, Dan, *Sir Banister Fletcher's a History of Architecture*, Architectural Press, 20th edition, 1996 (first published 1896). ISBN 0-7506-2267-9. Cf. Part One, Chapter 4.

- William W. Hallo & William Kelly Simpson, *The Ancient Near East: A History*, Holt Rinehart and Winston Publishers, 2nd edition, 1997. ISBN 0-15-503819-2.

- Jack Sasson, *The Civilizations of the Ancient Near East*, New York, 1995

- Marc Van de Mieroop, *History of the Ancient Near East: Ca. 3000-323 B.C.*, Blackwell Publishers, 2nd edition, 2006 (first published 2003). ISBN 1-4051-4911-6.

- Pittman, Holly (1984). Art of the Bronze Age: southeastern Iran, western Central Asia, and the Indus Valley. New York: The Metropolitan Museum of Art. ISBN 9780870993657. External link in |title= (help)

3.8 External links

- The History of the Ancient Near East – A database of the prehistoric Near East as well as its ancient history up to approximately the destruction of Jerusalem by the Romans ...

- Vicino Oriente – Vicino Oriente is the journal of the Section Near East of the Department of Historical, Archaeological and Anthropological Sciences of Antiquity of Rome 'La Sapienza' University. The Journal, which is published yearly, deals with Near Eastern History, Archaeology, Epigraphy, extending its view also on the whole Mediterranean with the study of Phoenician and Punic documents. It is accompanied by 'Quaderni di Vicino Oriente', a monograph series.

- Ancient Near East.net – an information and content portal for the archaeology, ancient history, and culture of the ancient Near East and Egypt

- Freer Gallery of Art, Smithsonian Institution The Freer Gallery houses a famous collection of ancient Near Eastern artefacts and records, notebooks and photographs of excavations in Samarra (Iraq), Persepolis and Pasargadae (Iran)

- The Freer Gallery of Art and Arthur M. Sackler Gallery Archives The archives for The Freer Gallery of Art and Arthur M. Sackler Gallery houses the papers of Ernst Herzfeld regarding his many excavations, along with records of other archeological excavations in the ancient Near East.

- Ancient Near East.org—a database of the prehistoric Near East as well as its ancient history up to approximately the destruction of Jerusalem by the Romans

- Archaeowiki.org—a wiki for the research and documentation of the ancient Near East and Egypt

- ETANA – website hosted by a consortium of universities in the interests of providing digitized resources and relevant web links

- Resources on Biblical Archaeology

- Ancient Near East Photographs This collection, created by Professor Scott Noegel, documents artifacts and archaeological sites of the ancient Near East; from the University of Washington Libraries Digital Image Collection

- Near East Images A directory of archaeological images of the ancient Near East

- Bioarchaeology of the Near East An Open Access journal

Chapter 4

History of Mesopotamia

The **history of Mesopotamia** describes the history of the area known as Mesopotamia, roughly coinciding with the Tigris–Euphrates basin, from the earliest human occupation in the Lower Palaeolithic period up to the Muslim conquests in the 7th century AD. This history is pieced together from evidence retrieved from archaeological excavations and, after the introduction of writing in the late 4th millennium BC, an increasing amount of historical sources. While in the Paleolithic and early Neolithic periods only parts of Upper Mesopotamia were occupied, the southern alluvium was settled during the late Neolithic period. Mesopotamia has been home to many of the oldest major civilizations, entering history from the Early Bronze Age, for which reason it is often dubbed the cradle of civilization. The rise of the first cities in southern Mesopotamia dates to the Chalcolithic (Uruk period), from c. 5300 BC; its regional independence ended with the Achaemenid conquest in 539 BC, although a few native neo-Assyrian kingdoms existed at different times, namely Adiabene, Osroene and Hatra.

4.1 Short outline of Mesopotamia

Main articles: Mesopotamia and Geography of Mesopotamia
 Mesopotamia literally means "(Land) between rivers" in ancient Greek. The oldest known occurrence of the name Mesopotamia dates to the 4th century BC, when it was used to designate the land east of the Euphrates in north Syria.[*][1] Later it was more generally applied to all the lands between the Euphrates and the Tigris, thereby incorporating not only parts of Syria but also almost all of Iraq and southeastern Turkey.[*][2] The neighbouring steppes to the west of the Euphrates and the western part of the Zagros Mountains are also often included under the wider term Mesopotamia.
A further distinction is usually made between Upper or Northern Mesopotamia and Lower or Southern Mesopotamia.[*][6] Upper Mesopotamia, also known as the Jezirah, is the area between the Euphrates and the Tigris from their sources down to Baghdad.[*][3] Lower Mesopotamia is the area from Baghdad to the Persian Gulf.[*][6] In modern scientific usage, the term Mesopotamia often also has a chronological connotation. It is usually used to designate the area until the Arab Muslim conquests in the 7th century AD, with Arabic names like Syria, Jezirah and Iraq being used to describe the region after that date.[*][2][*][7][*][nb 1]

4.1.1 Chronology and periodization

Further information: Chronology of the ancient Near East, ASPRO chronology and Dating methodologies in archaeology

Two types of chronologies can be distinguished: a relative chronology and an absolute chronology. The former establishes the order of phases, periods, cultures and reigns, whereas the latter establishes their absolute age expressed in years. In archaeology, relative chronologies are established by carefully excavating archaeological sites and reconstructing their stratigraphy – the order in which layers were deposited. In general, newer remains are deposited on top of older material. Absolute chronologies are established by dating remains, or the layers in which they are found, through absolute dating

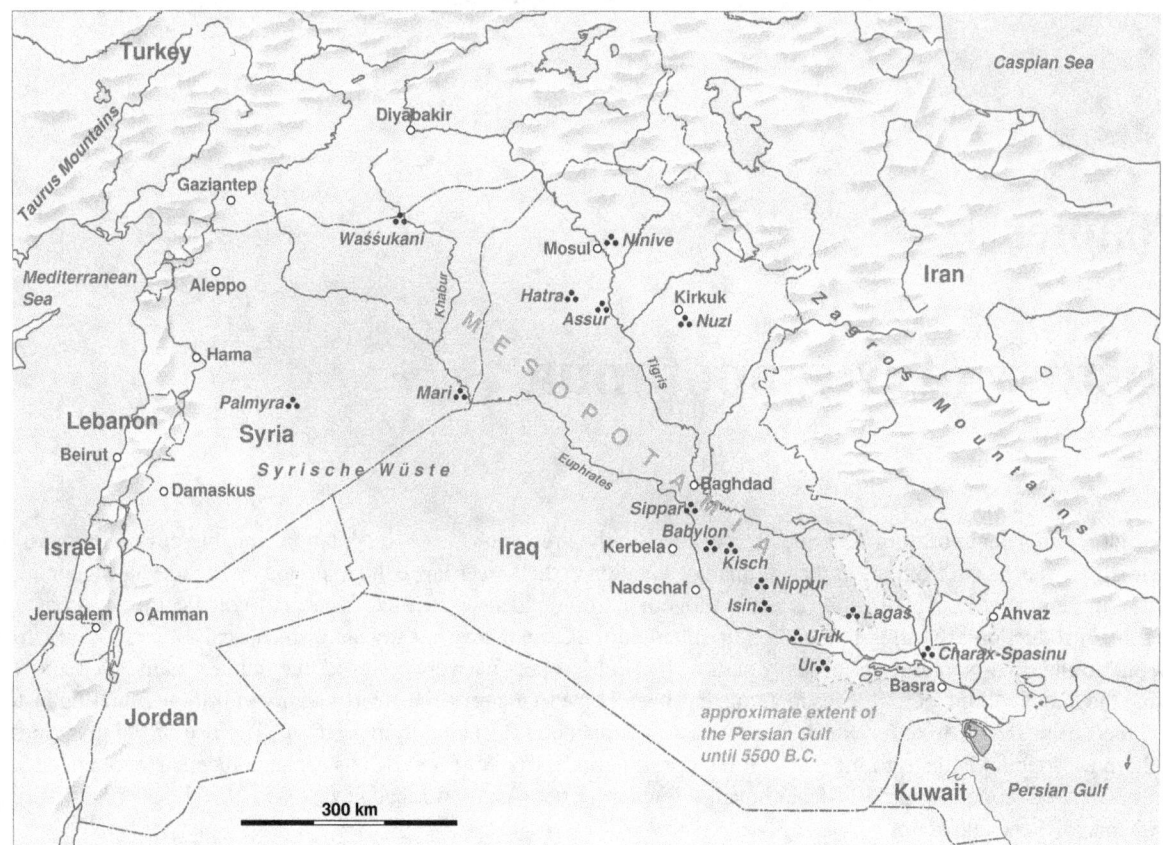

Map showing the extent of Mesopotamia

methods. These methods include radiocarbon dating and the written record that can provide year names or calendar dates. By combining absolute and relative dating methods, a chronological framework has been built for Mesopotamia that still incorporates many uncertainties but that also continues to be refined.[8][9] In this framework, many prehistorical and early historical periods have been defined on the basis of material culture that is thought to be representative for each period. These periods are often named after the site at which the material was recognized for the first time, as is for example the case for the Halaf, Ubaid and Jemdet Nasr periods.[8] When historical documents become widely available, periods tend to be named after the dominant dynasty or state; examples of this are the Ur III and Old Babylonian periods.[10] While reigns of kings can be securely dated for the 1st millennium BC, there is an increasingly large error margin toward the 2nd and 3rd millennia BC.[9]

Based on different estimates for the length of periods for which still very few historical documents are available, so-called Long, Middle, Short and Ultra-short Chronologies have been proposed by various scholars, varying by as much as 150 years in their dating of specific periods.[11][12] Despite problems with the Middle Chronology, this chronological framework continues to be used by many recent handbooks on the archaeology and history of the ancient Near East.[9][13][14][15][16] A study from 2001 published high-resolution radiocarbon dates from Turkey supporting dates for the 2nd millennium BC that are very close to those proposed by the Middle Chronology.[17][nb 2]

4.2 Prehistory

4.2.1 Pre-Pottery Neolithic period

The early Neolithic human occupation of Mesopotamia is, like the previous Epipaleolithic period, confined to the foothill zones of the Taurus and Zagros Mountains and the upper reaches of the Tigris and Euphrates valleys. The Pre-Pottery

Overview of Göbekli Tepe with modern roof to protect the site against the weather

Neolithic A (PPNA) period (10,000–8700 BC) saw the introduction of agriculture, while the oldest evidence for animal domestication dates to the transition from the PPNA to the Pre-Pottery Neolithic B (PPNB, 8700–6800 BC) at the end of the 9th millennium BC. This transition has been documented at sites like Abu Hureyra and Mureybet, which continued to be occupied from the Natufian well into the PPNB.[*18][*19] The so-far earliest monumental sculptures and circular stone buildings from Göbekli Tepe in southeastern Turkey date to the PPNA/Early PPNB and represent, according to the excavator, the communal efforts of a large community of hunter-gatherers.[*20][*21]

- Jarmo
- Samarra culture
- Halaf culture

4.2.2   Chalcolithic period

Main articles: Ubaid period and Uruk period

The Fertile Crescent was inhabited by several distinct, flourishing cultures between the end of the last ice age (c. 10,000 BC) and the beginning of history. One of the oldest known Neolithic sites in Mesopotamia is Jarmo, settled around 7000 BC and broadly contemporary with Jericho (in the Levant) and Çatal Hüyük (in Anatolia). It as well as other early Neolithic sites, such as Samarra and Tell Halaf were in northern Mesopotamia; later settlements in southern Mesopotamia required complicated irrigation methods. The first of these was Eridu, settled during the Ubaid period culture by farmers who brought with them the Samarran culture from the north. This was followed by Uruk period and the emergence of the Sumerians.

4.3   Third millennium BC

4.3.1   Jemdet Nasr period

Main article: Jemdet Nasr period

The Jemdet Nasr period, named after the type-site Jemdet Nasr, is generally dated to 3100–2900 BC.[*22] It was first distinguished on the basis of distinctive painted monochrome and polychrome pottery with geometric and figurative designs.[*23] The cuneiform writing system that had been developed during the preceding Uruk period was further refined. While the language in which these tablets were written cannot be identified with certainty for this period, it is thought to be Sumerian. The texts deal with administrative matters like the rationing of foodstuffs or lists of objects or animals.[24] Settlements during this period were highly organized around a central building that controlled all aspects of society. The economy focused on local agricultural production and sheep-and-goat pastoralism. The homogeneity of the Jemdet Nasr period across a large area of southern Mesopotamia indicates intensive contacts and trade between settlements. This is strengthened by the find of a sealing at Jemdet Nasr that lists a number of cities that can be identified, including Ur, Uruk and Larsa.[25]

4.3.2 Early Dynastic period

Main article: Early Dynastic Period (Mesopotamia)
 The entire Early Dynastic period is generally dated to 2900–2350 BC according to the Middle Chronology, or 2800–2230 BC according to the Short Chronology.[26] The Sumerians were firmly established in Mesopotamia by the middle of the 4th millennium BC, in the archaeological Uruk period, although scholars dispute when they arrived.[27] It is hard to tell where the Sumerians might have come from because the Sumerian language is a language isolate, unrelated to any other known language. Their mythology includes many references to the area of Mesopotamia but little clue regarding their place of origin, perhaps indicating that they had been there for a long time. The Sumerian language is identifiable from its initially logographic script which arose last half of the 4th millennium BC.

By the 3rd millennium BC, these urban centers had developed into increasingly complex societies. Irrigation and other means of exploiting food sources were being used to amass large surpluses. Huge building projects were being undertaken by rulers, and political organization was becoming ever more sophisticated. Throughout the millennium, the various city-states Kish, Uruk, Ur and Lagash vied for power and gained hegemony at various times. Nippur and Girsu were important religious centers, as was Eridu at this point. This was also the time of Gilgamesh, a semi-historical king of Uruk, and the subject of the famous *Epic of Gilgamesh*. By 2600 BC, the logographic script had developed into a decipherable cuneiform syllabic script.

The chronology of this era is particularly uncertain due to difficulties in our understanding of the text, our understanding of the material culture of the Early Dynastic period and a general lack of radiocarbon dates for sites in Iraq. Also, the multitude of city-states made for a confusing situation, as each had its own history. The Sumerian king list is one record of the political history of the period. It starts with mythological figures with improbably long reigns, but later rulers have been authenticated with archaeological evidence. The first of these is Enmebaragesi of Kish, c. 2600 BC, said by the king list to have subjected neighboring Elam. However, one complication of the Sumerian king list is that although dynasties are listed in sequential order, some of them actually ruled at the same time over different areas.

Enshakushanna of Uruk conquered all of Sumer, Akkad, and Hamazi, followed by Eannatum of Lagash who also conquered Sumer. His methods were force and intimidation (see the Stele of the Vultures), and soon after his death, the cities rebelled and the empire again fell apart. Some time later, Lugal-Anne-Mundu of Adab created the first, if short-lived, empire to extend west of Mesopotamia, at least according to historical accounts dated centuries later. The last native Sumerian to rule over most of Sumer before Sargon of Akkad established supremacy was Lugal-Zage-Si.

During the 3rd millennium BC, there developed a very intimate cultural symbiosis between the Sumerians and the Akkadians which included widespread bilingualism.[28] The influence of Sumerian on Akkadian (and vice versa) is evident in all areas, from lexical borrowing on a massive scale, to syntactic, morphological, and phonological convergence.[28] This has prompted scholars to refer to Sumerian and Akkadian in the 3rd millennium as a *sprachbund*.[28]

Akkadian gradually replaced Sumerian as the spoken language of Mesopotamia somewhere around the turn of the 3rd and the 2nd millennium BC (the exact dating being a matter of debate),[29] but Sumerian continued to be used as a sacred, ceremonial, literary and scientific language in Mesopotamia until the 1st century AD.

4.3.3 Akkadian Empire

Main article: Akkadian Empire
 The Akkadian period is generally dated to 2350–2170 BC according to the Middle Chronology, or 2230–2050 BC according to the Short Chronology.[26] Around 2334 BC, Sargon became ruler of Akkad in northern Mesopotamia. He proceeded to conquer an area stretching from the Persian Gulf into modern-day Syria. The Akkadians were a Semitic

people and the Akkadian language came into widespread use as the lingua franca during this period, but literacy remained in the Sumerian language. The Akkadians further developed the Sumerian irrigation system with the incorporation of large weirs and diversion dams into the design to facilitate the reservoirs and canals required to transport water vast distances.[30] The dynasty continued until around c. 2154 BC, and reached its zenith under Naram-Sin, who began the trend for rulers to claim divinity for themselves.

The Akkadian Empire lost power after the reign of Naram-Sin, and eventually was invaded by the Guti from the Zagros Mountains. For half a century the Guti controlled Mesopotamia, especially the south, but they left few inscriptions, so they are not well understood. The Guti hold loosened on southern Mesopotamia, where the second dynasty of Lagash came into prominence. Its most famous ruler was Gudea, who left many statues of himself in temples across Sumer.

4.3.4 Ur III period

Main article: Third Dynasty of Ur

Eventually the Guti were overthrown by Utu-hengal of Uruk, and the various city-states again vied for power. Power over the area finally went to the city-state of Ur, when Ur-Nammu founded the Ur III Empire (2112–2004 BC) and conquered the Sumerian region. Under his son Shulgi, state control over industry reached a level never again seen in the region. Shulgi may have devised the Code of Ur-Nammu, one of the earliest known law codes (three centuries before the more famous Code of Hammurabi). Around 2000 BC, the power of Ur waned, and the Amorites came to occupy much of the area, although it was Sumer's long-standing rivals to the east, the Elamites, who finally overthrew Ur. In the north, Assyria remained free of Amorite control until the very end of the 19th century BC. This marked the end of city-states ruling empires in Mesopotamia, and the end of Sumerian dominance, but the succeeding rulers adopted much of Sumerian civilization as their own.

4.4 Second millennium BC

4.4.1 Old Assyrian Period

Of the early history of the kingdom of Assyria, little is positively known. The Assyrian King List mentions rulers going back to the 23rd and 22nd century BC. The earliest king named Tudiya, who was a contemporary of Ibrium of Ebla, appears to have lived in the mid-23rd century BC, according to the king list. Tudiya concluded a treaty with Ibrium for the use of a trading post in The Levant officially controlled by Ebla. Apart from this reference to trading activity, nothing further has yet been discovered about Tudiya. He was succeeded by Adamu and then a further thirteen rulers about all of whom nothing is yet known. These early kings from the 23rd to late 21st centuries BC, who are recorded as *kings who lived in tents* were likely to have been semi nomadic pastoralist rulers, nominally independent but subject to the Akkadian Empire, who dominated the region and at some point during this period became fully urbanised and founded the *city state* of Ashur.[31] A king named Ushpia (c. 2030 BC) is credited with dedicating temples to Ashur in the home city of the god. In around 1975 BC Puzur-Ashur I founded a new dynasty, and his successors such as Shalim-ahum, Ilushuma (1945–1906 BC), Erishum I (1905–1867 BC), Ikunum (1867–1860 BC), Sargon I, Naram-Sin and Puzur-Ashur II left inscriptions regarding the building of temples to Ashur, Adad and Ishtar in Assyria. Ilushuma in particular appears to have been a powerful king and the dominant ruler in the region, who made many raids into southern Mesopotamia between 1945 BC and 1906 BC, attacking the independent Sumero-Akkadian city states of the region such as Isin, and founding colonies in Asia Minor. This was to become a pattern throughout the history of ancient Mesopotamia with the future rivalry between Assyria and Babylonia. However, Babylonia did not exist at this time, but was founded in 1894 BC by an Amorite prince named Sumuabum during the reign of Erishum I.

4.4.2 Isin-Larsa, Old Babylonian and Shamshi-Adad I

Further information: First Babylonian Dynasty

The next two centuries or so saw southern Mesopotamia dominated by the Amorite cities of Isin and Larsa, as the two cities vied for dominance. This period also marked a growth in power in the north of Mesopotamia. An Assyrian king named Ilushuma (1945–1906 BC) became a dominant figure in Mesopotamia, raiding the southern city states and founding colonies in Asia Minor. Eshnunna and Mari, two Amorite ruled states also became important in the north.

Babylonia was founded as an independent state by an Amorite chieftain named Sumuabum in 1894 BC. For over a century after its founding, it was a minor and relatively weak state, overshadowed by older and more powerful states such as Isin, Larsa, Assyria and Elam. However, Hammurabi (1792 BC to 1750 BC), the Amorite ruler of Babylon, turned Babylon into a major power and eventually conquered Mesopotamia and beyond. He is famous for his law code and conquests, but he is also famous due to the large amount of records that exist from the period of his reign. After the death of Hammurabi, the first Babylonian dynasty lasted for another century and a half, but his empire quickly unravelled, and Babylon once more became a small state. The Amorite dynasty ended in 1595 BC, when Babylonia fell to the Hittite king Mursilis, after which the Kassites took control.

Unlike the south of Mesopotamia, the native Akkadian kings of Assyria repelled Amorite advances during the 20th and 19th centuries BC. However this changed in 1813 BC when an Amorite king named Shamshi-Adad I usurped the throne of Assyria. Although claiming descendency from the native Assyrian king Ushpia, he was regarded as an interloper. Shamshi-Adad I created a regional empire in Assyria, maintaining and expanding the established colonies in Asia Minor and Syria. His son Ishme-Dagan I continued this process, however his successors were eventually conquered by Hammurabi, a fellow Amorite from Babylon. The three Amorite kings succeeding Ishme-Dagan were vassals of Hammurabi, but after his death, a native Akkadian vice regent Puzur-Sin overthrew the Amorites of Babylon and a period of civil war with multiple claimants to the throne ensued, ending with the succession of king Adasi c. 1720 BC.

4.4.3 Middle Assyrian Period and Empire

The Middle Assyrian period begins c. 1720 BC with the ejection of Amorites and Babylonians from Assyria by a king called Adasi. The nation remained relatively strong and stable, peace was made with the Kassite rulers of Babylonia, and Assyria was free from Hittite, Hurrian, Gutian, Elamite and Mitanni threat. However a period of Mitanni domination occurred from the mid-15th to early 14th centuries BC. This was ended by Eriba-Adad I (1392 BC - 1366), and his successor Ashur-uballit I completely overthrew the Mitanni Empire and founded a powerful Assyrian Empire that came to dominate Mesopotamia and much of the ancient Near East (including Babylonia, Asia Minor, Iran, the Levant and parts of the Caucasus and Arabia), with Assyrian armies campaigning from the Mediterranean Sea to the Caspian, and from the Caucasus to Arabia. The empire endured until 1076 BC with the death of Tiglath-Pileser I. During this period Assyria became a major power, overthrowing the Mitanni Empire, annexing swathes of Hittite, Hurrian and Amorite land, sacking and dominating Babylon, Canaan/Phoenicia and becoming a rival to Egypt.

4.4.4 Kassite dynasty of Babylon

Main article: Kassites

Although the Hittites overthrew Babylon, another people, the Kassites, took it as their capital (c. 1650–1155 BC (short chronology)). They have the distinction of being the longest lasting dynasty in Babylon, reigning for over four centuries. They left few records, so this period is unfortunately obscure. They are of unknown origin; what little we have of their language suggests it is a language isolate. Although Babylonia maintained its independence through this period, it was not a power in the Near East, and mostly sat out the large wars fought over the Levant between Egypt, the Hittite Empire, and Mitanni (see below), as well as independent peoples in the region. Assyria participated in these wars toward the end of the period, overthrowing the Mitanni Empire and besting the Hittites and Phrygians, but the Kassites in Babylon did not. They did, however, fight against their longstanding rival to the east, Elam (related by some linguists to the Dravidian languages in modern India). Babylonia found itself under Assyrian and Elamite domination for much of the later Kassite period. In the end, the Elamites conquered Babylon, bringing this period to an end.

4.4.5 Hurrians

Main article: Mitanni

The Hurrians were a people who settled in north western Mesopotamia and South-East Anatolia in 1600 BC. By 1450 BC they established a medium-sized empire under a Mitanni ruling class, and temporarily made tributary vassals out of kings in the west, making them a major threat for the Pharaoh in Egypt until their overthrow by Assyria. The Hurrian language is related to the later Urartian, but there is no conclusive evidence these two languages are related to any others.

4.4.6 Hittites

Main article: Hittites

By 1300 BC the Hurrians had been reduced to their homelands in Asia Minor after their power was broken by the Assyrians and Hittites, and held the status of vassals to the "Hatti", the Hittites, a western Indo-European people (belonging to the linguistic "kentum" group) who dominated most of Asia Minor (modern Turkey) at this time from their capital of Hattusa. The Hittites came into conflict with the Assyrians from the mid-14th to the 13th centuries BC, losing territory to the Assyrian kings of the period. However they endured until being finally swept aside by the Phrygians, who conquered their homelands in Asia Minor. The Phrygians were prevented from moving south into Mesopotamia by the Assyrian king Tiglath-Pileser I. The Hittites fragmented into a number of small Neo-Hittite states, which endured in the region for many centuries.

4.4.7 Bronze Age collapse

Main article: Bronze Age collapse

Records from the 12th and 11th centuries BC are sparse in Babylonia, which had been overrun with new Semitic settlers, namely the Arameans, Chaldeans and Sutu. Assyria however, remained a compact and strong nation, which continued to provide much written record. The 10th century BC is even worse for Babylonia, with very few inscriptions. Mesopotamia was not alone in this obscurity: the Hittite Empire fell at the beginning of this period and very few records are known from Egypt and Elam. This was a time of invasion and upheaval by many new people throughout the Near East, North Africa, The Caucasus, Mediterranean and Balkan regions.

4.5 First millennium BC

4.5.1 Neo-Assyrian Empire

Main article: Neo-Assyrian Empire

The Neo-Assyrian Empire is usually considered to have begun with the accession of Adad-nirari II, in 911 BC, lasting until the fall of Nineveh at the hands of the Babylonians, Medes, Scythians and Cimmerians in 612 BC. The empire was the largest and most powerful the world had yet seen. At its height Assyria conquered the 25th dynasty Egypt (and expelled its Nubian/Kushite dynasty) as well as Babylonia, Chaldea, Elam, Media, Persia, Urartu, Phoenicia, Aramea/Syria, Phrygia, the Neo-Hittites, Hurrians, northern Arabia, Gutium, Israel, Judah, Moab, Edom, Corduene, Cilicia, Mannea and parts of Ancient Greece (such as Cyprus), and defeated and/or exacted tribute from Scythia, Cimmeria, Lydia, Nubia, Ethiopia and others.

4.5.2   Neo-Babylonian Empire

Main article: Neo-Babylonian Empire

The Neo-Babylonian Empire or Second Babylonian Empire was a period of Mesopotamian history which began in 620 BC and ended in 539 BC. During the preceding three centuries, Babylonia had been ruled by their fellow Akkadian speakers and northern neighbours, Assyria. The Assyrians had managed to maintain Babylonian loyalty through the Neo-Assyrian period, whether through granting of increased privileges, or militarily, but that finally changed after 627 BC with the death of the last strong Assyrian ruler, Ashurbanipal, and Babylonia rebelled under Nabopolassar a Chaldean chieftain the following year. In alliance with king Cyaxares of the Medes,and with the help of the Scythians and Cimmerians the city of Nineveh was sacked in 612 BC, Assyria fell by 605 BC and the seat of empire was transferred to Babylonia for the first time since Hammurabi.

4.5.3   Classical Antiquity to Late Antiquity

After the death of Ashurbanipal in 627 BC, the Assyrian empire descended into a series of bitter civil wars, allowing its former vassals to free themselves. Cyaxares reorganized and modernized the Median Army, then joined with King Nabopolassar of Babylon. These allies, together with the Scythians, overthrew the Assyrian Empire and destroyed Nineveh in 612 BC. After the final victory at Carchemish in 605 BC the Medes and Babylonians ruled Assyria. Babylon and Media fell under Persian rule in the 6th century BC (Cyrus the Great).

For two centuries of Achaemenid rule both Assyria and Babylonia flourished, Achaemenid Assyria in particular becoming a major source of manpower for the army and a breadbasket for the economy. Mesopotamian Aramaic remained the *lingua franca* of the Achaemenid Empire, much as it had done in Assyrian times. Mesopotamia fell to Alexander the Great in 330 BC, and remained under Hellenistic rule for another two centuries, with Seleucia as capital from 305 BC. In the 1st century BC, Mesopotamia was in constant turmoil as the Seleucid Empire was weakened by Parthia on one hand and the Mithridatic Wars on the other. The Parthian Empire lasted for five centuries, into the 3rd century AD, when it was succeeded by the Sassanids. Christianity entered Mesopotamia from the 1st to 3rd centuries AD, and flourished, particularly in Assyria (Assuristan in Sassanid Persian), which became the center of the Assyrian Church of the East and a flourishing Syriac Christian tradition which remains to this day. A number of Neo-Assyrian kingdoms arose, in particular Adiabene. The Sassanid Empire finally fell to the Rashidun army under Khalid ibn al-Walid in the 630s. After the Arab-Islamic conquest of the mid-7th century AD, Mesopotamia saw an influx of non native Arabs and later also Turkic peoples, this and the spread of Islam and Arabic led to the gradual marginalisation of native Mesopotamians. However, a number of Assyrians (known as Ashuriyun by the Arabs) resisted the Arabisation and Islamification of Mesopotamia. The city of Assur was still occupied until the 14th century, and Assyrians possibly still formed the majority in northern Mesopotamia until the Middle Ages. Assyrians retain Eastern Rite Christianity and Mesopotamian Aramaic as a mother tongue and written script to this day. Among these people, the giving of traditional Mesopotamian names is still common.

- Classical Antiquity
  - Median and Babylonian Assyria (605 to 549 BC)
  - Persian Babylonia,
  - Achaemenid Assyria (6th to 4th centuries BC)
  - Seleucid Mesopotamia (4th to 3rd centuries BC)
  - Parthian Asuristan (Assyria (3rd century BC to 3rd century AD)
  - Osroene (2nd century BC to 3rd century AD)
  - Adiabene (1st to 2nd centuries AD)
  - Roman Mesopotamia, Roman Assyria (2nd century AD)

- Late Antiquity
  - Sassanid Asuristan (Assyria) (3rd to 7th centuries AD)
  - Arab Muslim conquest of Mesopotamia, dissolution of Assyria (633–651 AD)

4.6 See also

- History of Iraq
- History of Syria

4.7 Notes

[1] This page will use Mesopotamia in its widest geographical and chronological sense.

[2] This page will use the Middle Chronology.

4.8 References

[1] Finkelstein 1962, p. 73

[2] Foster & Polinger Foster 2009, p. 6

[3] Canard 2011

[4] Wilkinson 2000, pp. 222–223

[5] Matthews 2003, p. 5

[6] Miquel et al. 2011

[7] Bahrani 1998

[8] Matthews 2003, pp. 65–66

[9] van de Mieroop 2007, p. 4

[10] van de Mieroop 2007, p. 3

[11] Brinkman 1977

[12] Gasche et al. 1998

[13] Kuhrt 1997, p. 12

[14] Potts 1999, p. xxix

[15] Akkermans & Schwartz 2003, p. 13

[16] Sagona & Zimansky 2009, p. 251

[17] Manning et al. 2001

[18] Moore, Hillman & Legge 2000

[19] Akkermans & Schwartz 2003

[20] Schmidt 2003

[21] Banning 2011

[22] Pollock 1999, p. 2

[23] Matthews 2002, pp. 20–21

[24] Woods 2010, pp. 36–45

[25] Matthews 2002, pp. 33–37

[26] Pruß 2004

[27] Woolley 1965, p. 9

[28] Deutscher 2007, pp. 20–21

[29] Woods 2006

[30] http://mygeologypage.ucdavis.edu/cowen/~{}GEL115/115CH17oldirrigation.html

[31] Saggs, *The Might*, 24.

4.9   Bibliography

- Akkermans, Peter M.M.G.; Schwartz, Glenn M. (2003). *The Archaeology of Syria. From Complex Hunter-Gatherers to Early Urban Societies (ca. 16,000–300 BC)*. Cambridge: Cambridge University Press. ISBN 0-521-79666-0.

- Bahrani, Z. (1998). "Conjuring Mesopotamia: Imaginative Geography a World Past". In Meskell, L. *Archaeology under Fire: Nationalism, Politics and Heritage in the Eastern Mediterranean and Middle East*. London: Routledge. pp. 159–174. ISBN 978-0-415-19655-0.

- Banning, E.B. (2011).   "So Fair a House. Göbekli Tepe and the Identification of Temples in the Pre-Pottery Neolithic of the Near East". *Current Anthropology* **52** (5): 619–660. doi:10.1086/661207.

- Braidwood, Robert J.; Howe, Bruce (1960). *Prehistoric Investigations in Iraqi Kurdistan* (PDF). Studies in Ancient Oriental Civilization **31**. Chicago: University of Chicago Press. OCLC 395172.

- Brinkman, J.A. (1977).   "Appendix: Mesopotamian Chronology of the Historical Period". In Oppenheim, A.L. *Ancient Mesopotamia: Portrait of a Dead Civilization*. Chicago: University of Chicago Press. pp. 335–348. ISBN 0-226-63186-9.

- Canard, M. (2011).   "al-D̲jazīra, D̲jazīrat Aḳūr or Iḳlīm Aḳūr". In Bearman, P.; Bianquis, Th.; Bosworth, C.E.; van Donzel, E.; Heinrichs, W.P. *Encyclopaedia of Islam, Second Edition*. Leiden: Brill Online. OCLC 624382576.

- Deutscher, Guy (2007). *Syntactic Change in Akkadian: The Evolution of Sentential Complementation*. Oxford University Press. ISBN 978-0-19-953222-3.

- Finkelstein, J.J. (1962).   "Mesopotamia". *Journal of Near Eastern Studies* **21** (2): 73–92. doi:10.1086/371676. JSTOR 543884.

- Foster, Benjamin R.; Polinger Foster, Karen (2009). *Civilizations of Ancient Iraq*. Princeton: Princeton University Press. ISBN 978-0-691-13722-3.

- Gasche, H.; Armstrong, J.A.; Cole, S.W.; Gurzadyan, V.G. (1998). *Dating the Fall of Babylon: a Reappraisal of Second-Millennium Chronology*. Chicago: University of Ghent/Oriental Institute of the University of Chicago. ISBN 1-885923-10-4.

- Hunt, Will (2010). *Arbil, Iraq Discovery Could be Earliest Evidence of Humans in the Near East*. Heritage Key. Retrieved 4 August 2010.

- Kozłowski, Stefan Karol (1998).   "M'lefaat: Early Neolithic Site in Northern Irak". *Cahiers de l'Euphrate* **8**: 179–273. OCLC 468390039.

- Kuhrt, A. (1997). *Ancient Near East c. 3000–330 BC*. London: Routledge. ISBN 0-415-16763-9.

- Manning, S.W.; Kromer, B.; Kuniholm, P.I.; Newton, M.W. (2001). *Anatolian Tree Rings and a New Chronology for the East Mediterranean Bronze-Iron Ages*. *Science* **294** (5551). pp. 2532–2535. doi:10.1126/science.1066112. PMID 11743159.

- Matthews, Roger (2002). *Secrets of the Dark Mound: Jemdet Nasr 1926–1928*. Iraq Archaeological Reports **6**. Warminster: BSAI. ISBN 0-85668-735-9.

- Matthews, Roger (2003). *The Archaeology of Mesopotamia: Theories and Approaches*. Approaching the past. Milton Square: Routledge. ISBN 0-415-25317-9.

- Miquel, A.; Brice, W.C.; Sourdel, D.; Aubin, J.; Holt, P.M.; Kelidar, A.; Blanc, H.; MacKenzie, D.N.; Pellat, Ch. (2011). "ʿIrāḳ". In Bearman, P.; Bianquis, Th.; Bosworth, C.E.; van Donzel, E.; Heinrichs, W.P. *Encyclopaedia of Islam, Second Edition*. Leiden: Brill Online. OCLC 624382576.

- Mohammadifar, Yaghoub; Motarjem, Abbass (2008). "Settlement Continuity in Kurdistan". *Antiquity* **82** (317). ISSN 0003-598X.

- Moore, A.M.T.; Hillman, G.C.; Legge, A.J. (2000). *Village on the Euphrates: From Foraging to Farming at Abu Hureyra*. Oxford: Oxford University Press. ISBN 0-19-510807-8.

- Muhesen, Sultan (2002). "The Earliest Paleolithic Occupation in Syria". In Akazawa, Takeru; Aoki, Kenichi; Bar-Yosef, Ofer. *Neandertals and Modern Humans in Western Asia*. New York: Kluwer. pp. 95–105. doi:10.1007/0-306-47153-1_7. ISBN 0-306-47153-1.

- Pollock, Susan (1999). *Ancient Mesopotamia: The Eden That Never Was*. Case Studies in Early Societies. Cambridge: Cambridge University Press. ISBN 978-0-521-57568-3.

- Potts, D.T. (1999). *The Archaeology of Elam*. Cambridge: Cambridge University Press. ISBN 0-521-56496-4.

- Pruß, Alexander (2004), "Remarks on the Chronological Periods", in Lebeau, Marc; Sauvage, Martin, *Atlas of Preclassical Upper Mesopotamia*, Subartu **13**, pp. 7–21, ISBN 2503991203

- Sagona, A.; Zimansky, P. (2009). *Ancient Turkey*. London: Routledge. ISBN 0-415-28916-5.

- Schmid, P.; Rentzel, Ph.; Renault-Miskovsky, J.; Muhesen, Sultan; Morel, Ph.; Le Tensorer, Jean Marie; Jagher, R. (1997). *Découvertes de Restes Humains dans les Niveaux Acheuléens de Nadaouiyeh Aïn Askar (El Kowm, Syrie Centrale)*. *Paléorient* **23** (1). pp. 87–93. doi:10.3406/paleo.1997.4646.

- Schmidt, Klaus (2003). *The 2003 Campaign at Göbekli Tepe (Southeastern Turkey)* (PDF). *Neo-Lithics* **2/03**. pp. 3–8. ISSN 1434-6990. Retrieved 21 October 2011.

- Solecki, Ralph S. (1997). "Shanidar Cave". In Meyers, Eric M. *The Oxford Encyclopedia of Archaeology in the Ancient Near East* **5**. New York: Oxford University Press. pp. 15–16. ISBN 0-19-506512-3.

- van de Mieroop, M. (2007). *A History of the Ancient Near East, ca. 3000–323 BC*. Malden: Blackwell. ISBN 0-631-22552-8.

- Wilkinson, Tony J. (2000). *Regional Approaches to Mesopotamian Archaeology: the Contribution of Archaeological Surveys*. *Journal of Archaeological Research* **8** (3). pp. 219–267. doi:10.1023/A:1009487620969. ISSN 1573-7756.

- Woods, Christopher (2006). "Bilingualism, Scribal Learning, and the Death of Sumerian". In Sanders, S.L. *Margins of Writing, Origins of Culture* (PDF). Chicago: University of Chicago Press. pp. 91–120.

- Woods, Christopher (2010). "The Earliest Mesopotamian Writing". In Woods, Christopher. *Visible Language: Inventions of Writing in the Ancient Middle East and Beyond* (PDF). Oriental Institute Museum Publications **32**. Chicago: University of Chicago. pp. 33–50. ISBN 978-1-885923-76-9.

- Woolley, C.L. (1965). *The Sumerians*. New York: W.W. Norton.

4.10   Further reading

- Joannès, Francis (2004). *The Age of Empires: Mesopotamia in the First Millennium BC*. Edinburgh: Edinburgh University Press. ISBN 0-7486-1755-8.

- Matthews, Roger (2000). *The Early Prehistory of Mesopotamia: 500,000 to 4,500 BC*. Subartu **5**. Turnhout: Brepols. ISBN 2-503-50729-8.

- Nissen, Hans J. (1988). *The Early History of the Ancient Near East 9000–2000 B.C.* London: University of Chicago Press. ISBN 0-226-58656-1.

- Postgate, J.N. (1992). *Early Mesopotamia: Society and Economy at the Dawn of History*. London: Routledge. ISBN 978-0-415-11032-7.

- Saggs, H.W.F. (1990). *Assyria: The Might that Was*. London: Sidgwick and Johnson. ISBN 0-312-03511-X.

- Simpson, St. John (1997). "Mesopotamia from Alexander to the Rise of Islam". In Meyers, Eric M. *The Oxford Encyclopedia of Archaeology in the Ancient Near East* **3**. New York: Oxford University Press. pp. 484–487. ISBN 0-19-506512-3.

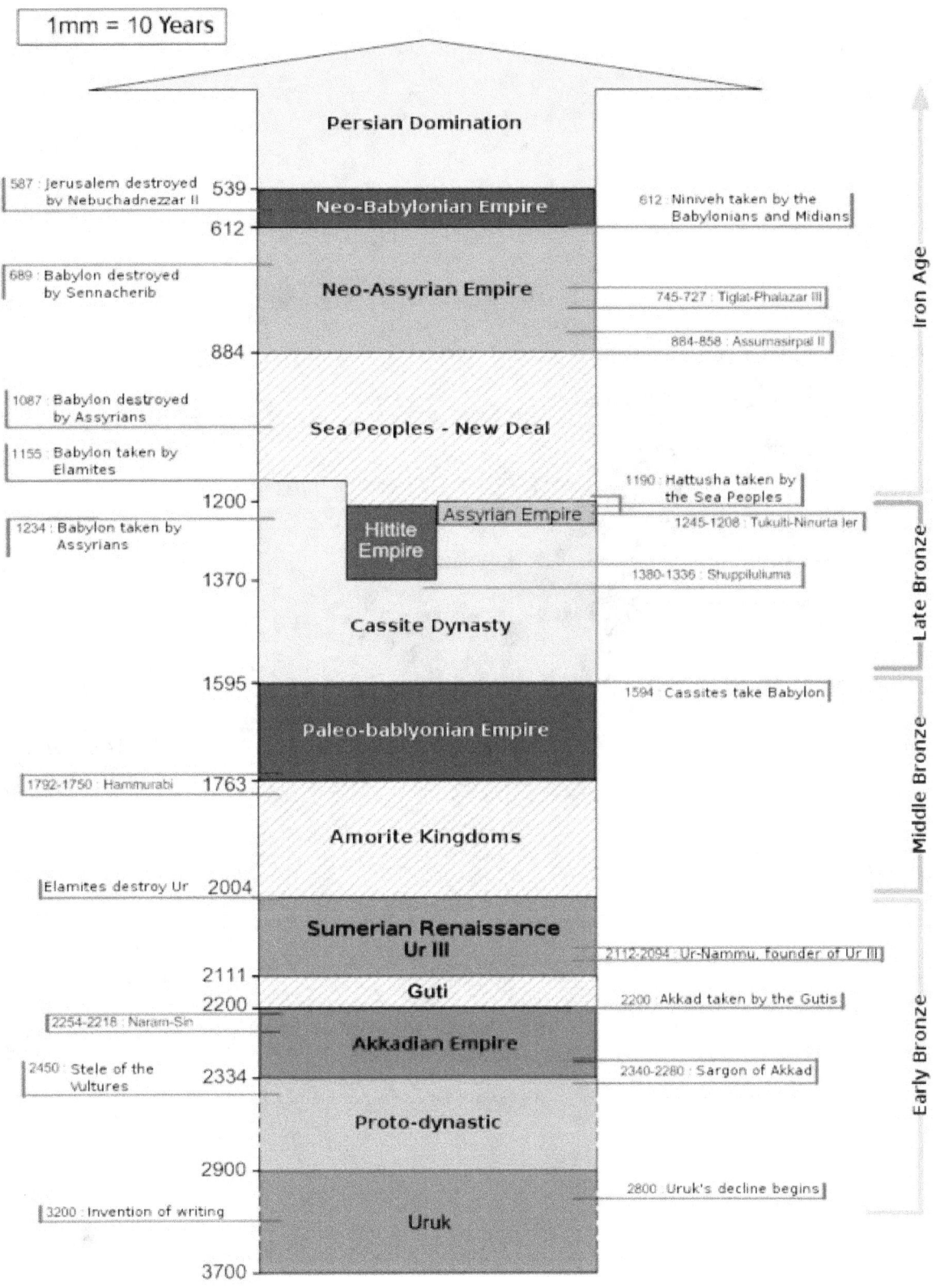

1mm = 10 Years

Persian Domination

587 : Jerusalem destroyed by Nebuchadnezzar II — 539

Neo-Babylonian Empire

612 : Niniveh taken by the Babylonians and Midians

612

689 : Babylon destroyed by Sennacherib

Neo-Assyrian Empire

745-727 : Tiglat-Phalazar III

884-858 : Assurnasirpal II

884

1087 : Babylon destroyed by Assyrians

Sea Peoples - New Deal

1155 : Babylon taken by Elamites

1190 : Hattusha taken by the Sea Peoples

1200

1234 : Babylon taken by Assyrians

Assyrian Empire

Hittite Empire

1245-1208 : Tukulti-Ninurta Ier

1380-1336 : Shuppiluliuma

1370

Cassite Dynasty

1595

1594 : Cassites take Babylon

Paleo-bablyonian Empire

1792-1750 : Hammurabi — 1763

Amorite Kingdoms

Elamites destroy Ur — 2004

Sumerian Renaissance Ur III

2111

2112-2094 : Ur-Nammu, founder of Ur III

Guti

2200

2200 : Akkad taken by the Gutis

2254-2218 : Naram-Sin

Akkadian Empire

2450 : Stele of the Vultures — 2334

2340-2280 : Sargon of Akkad

Proto-dynastic

2900

2800 : Uruk's decline begins

3200 : Invention of writing

Uruk

3700

Iron Age

Late Bronze

Middle Bronze

Early Bronze

Chronology of the main dominations

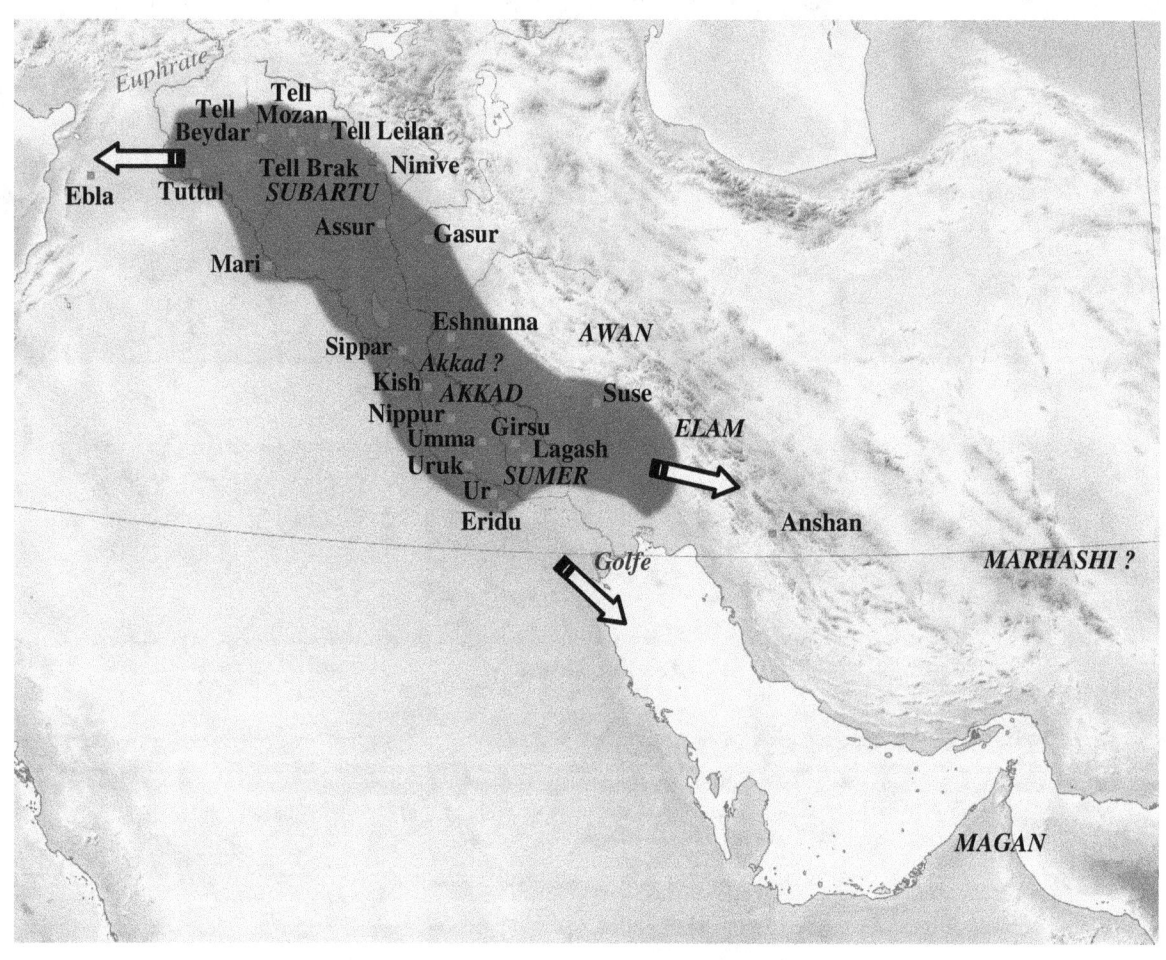

Map of the Akkadian Empire (brown) and the directions in which military campaigns were conducted (yellow arrows)

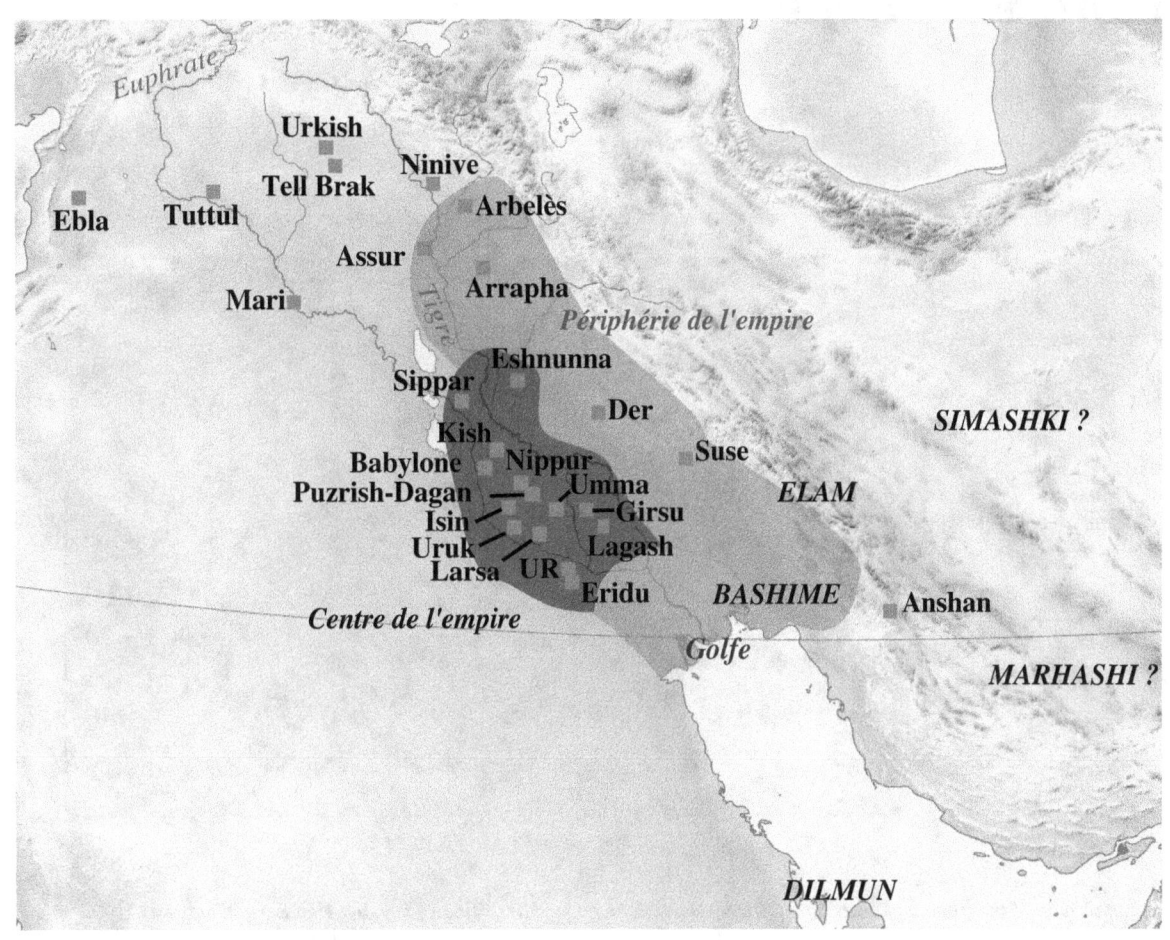

Map of the Ur III state (brown) and its influence sphere (red)

Chapter 5

Iranian Plateau

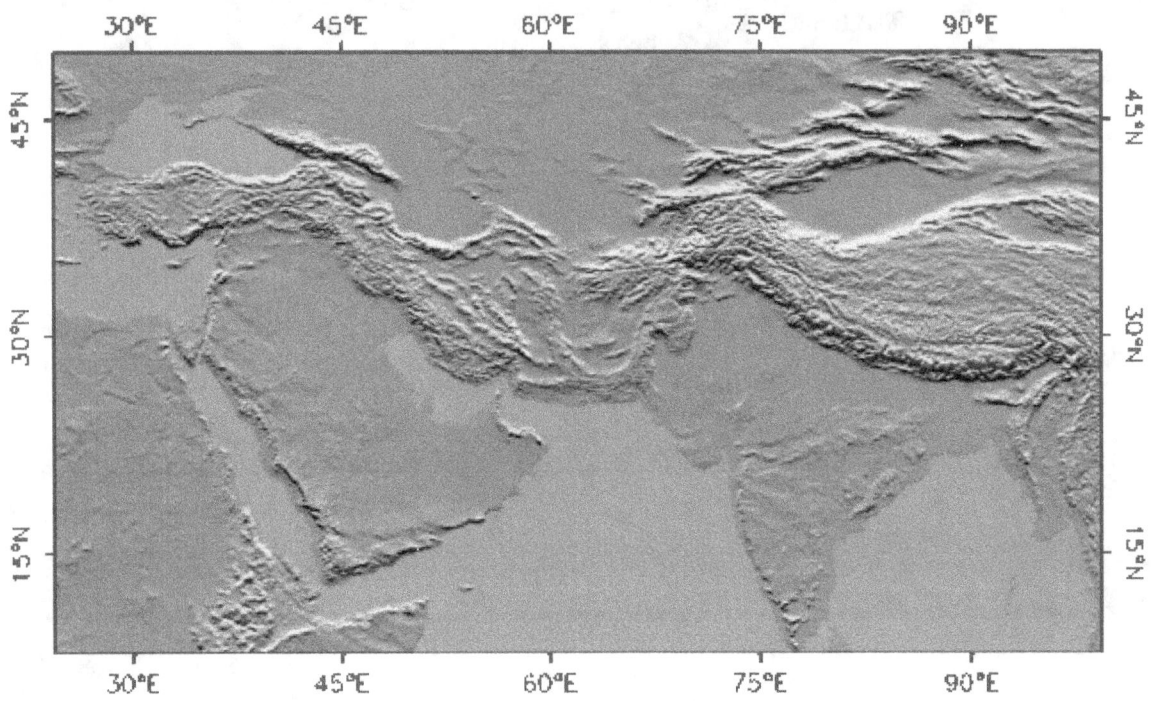

Topographic map of the Iranian plateau connecting to Anatolia in the west and Hindu Kush and Himalaya in the east. The Caspian Sea is absent from the topographic rendering. It is located in what appears here as the flatland depression directly north of the Persian Gulf at the border of the Alborz mountain range.

The **Persian Plateau**,[*][1][*][2] or **Iranian Plateau**, is a geological formation in Western Asia and Central Asia. It is the part of the Eurasian Plate wedged between the Arabian and Indian plates, situated between the Zagros Mountains to the west, the Caspian Sea and the Kopet Dag to the north, the Armenian Highlands and the Caucasus mountains in the northwest, the Hormuz Strait and Persian gulf to the south and the Indus River to the east in Pakistan.

As a historical region, it includes Parthia, Media, Persis, the heartlands of Iran and some of its recently lost territories.[*][3] The Zagros Mountains form the plateau's western boundary, and its eastern slopes may be included in the term. The Encyclopædia Britannica excludes "lowland Khuzestan" explicitly[*][4] and characterizes Elam as spanning "the region from the Mesopotamian plain to the Iranian Plateau".[*][5]

From the Caspian in the northwest to Baluchistan in the south-east, the Iranian Plateau extends for close to 2,000 km. It encompasses the greater part of Iran, Afghanistan and Pakistan west of the Indus River on an area roughly outlined by

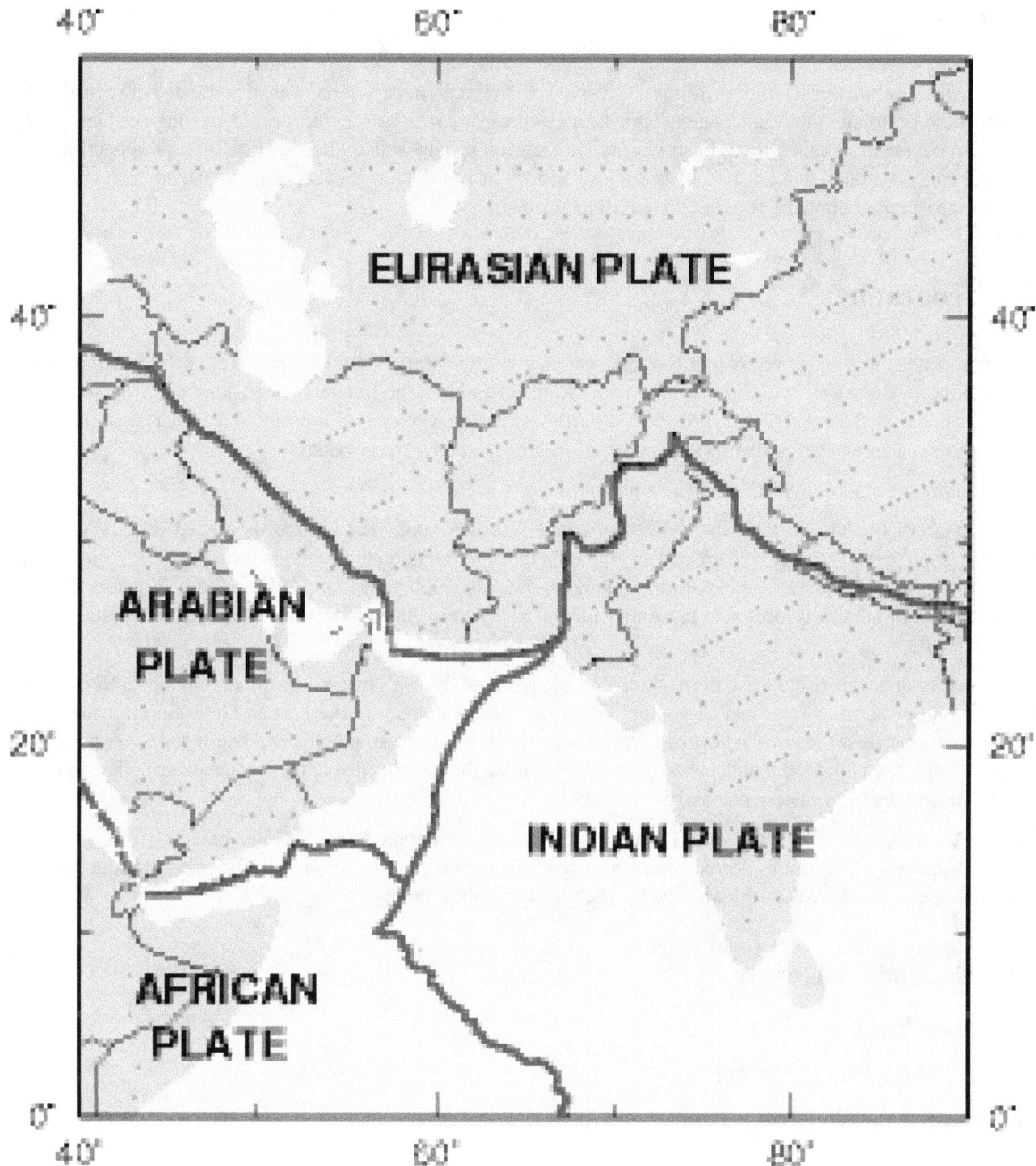

Closeup of the boundaries with the Eurasian, Arabian and Indian plates.

the quadrangle formed by the cities of Tabriz, Shiraz, Peshawar and Quetta containing some 3,700,000 square kilometres (1,400,000 sq mi). In spite of being called a "plateau", it is far from flat but contains several mountain ranges, the highest peak being Damavand in the Alborz at 5610 m, and the Lut basin east of Kerman in Central Iran falling below 300 m.

5.1 Geology

In geology, the plateau region of Iran primarily formed of the accretionary Gondwanan terranes between the Turan platform to the north and the Main Zagros Thrust, the suture zone between the northward moving Arabian plate and the Eurasian continent, is called the Iranian plateau. It is a geologically well-studied area because of general interest in continental collision zones, and because of Iran's long history of research in geology, particularly in economic geology (although Iran's major petroleum reserves are not in the plateau).

5.2 Geography

The Iranian plateau in geology refers to a geographical area north of the great folded mountain belts resulting from the collision of the Arabian plate with the Eurasian plate. In this definition, the Iranian plateau does not cover southwestern Iran. It extends from East Azerbaijan Province in northwest of Iran (Persia) all the way to Pakistan west of the Indus River. It also includes smaller parts of the Republic of Azerbaijan and Turkmenistan.

Its mountain ranges can be divided into five major sub-regions*[6] (see below).

The Northwestern Iranian Plateau, where the Pontic and Taurus Mountains converge, is rugged country with higher elevations, a more severe climate, and greater precipitation than are found on the Anatolian Plateau. The region is known as the Anti-Taurus, and the average elevation of its peaks exceeds 3,000 m. Mount Ararat, at 5,137 meters (16,854 ft) the highest point in Turkey, is located in the Anti-Taurus. Lake Van is situated in the mountains at an elevation of 1,546 meters (5,072 ft).

The headwaters of major rivers arise in the Anti-Taurus: the east-flowing Aras River, which empties into the Caspian Sea; the south-flowing Euphrates and Tigris join in Iraq before emptying into the Persian Gulf. Several small streams that empty into the Black Sea or landlocked Lake Van also originate in these mountains. The Indus River begins in the highlands of Tibet and flows the length of Pakistan almost tracing the eastern edge of the Iranian plateau. The Indus River forms the Iranian plateau's eastern boundary.

Southeast Anatolia lies south of the Anti-Taurus Mountains. It is a region of rolling hills and a broad plateau surface that extends into Syria. Elevations decrease gradually, from about 800 meters (2,600 ft) in the north to about 500 meters (1,600 ft) in the south. Traditionally, wheat and barley are the main crops of the region.

5.2.1 Mountain ranges

- Northwest Iran Ranges

    - Sabalan 4,811 m (15,784 ft)

- Alborz

    - Damavand 5,610 m (18,410 ft)

- Central Iranian Plateau

    - Kūh-e Hazār 4,500 m (14,800 ft)
    - Kuh-e Jebal Barez

- Eastern Iranian Ranges

    - Kopet Dag
        - Kuh-e Siah Khvani 3,314 m (10,873 ft) 36°17′N 59°3′E / 36.283°N 59.050°E
    - Eshdeger Range
        - 2,920 m (9,580 ft) 33°32′N 57°14′E / 33.533°N 57.233°E

- Balochistan

- Sikaram 4,755 m (15,600 ft) 34°2′N 69°54′E / 34.033°N 69.900°E

- Kuh-e Taftan 3,941 m (12,930 ft) 28°36′N 61°8′E / 28.600°N 61.133°E

- Zargun 3,578 m (11,739 ft) 30°16′N 67°18′E / 30.267°N 67.300°E

5.2.2 Rivers and plains

- Kavir Desert

- Lut Desert

- Hamun-e Jaz Murian

  - Halil River

- Gavkhouni

  - Zayandeh River

- Sistan Basin

  - Helmand River

  - Farah River

5.3 History

Main articles: Greater Iran and History of Iran
Further information: Airyanem Vaejah

In the Bronze Age, Elam stretched across the Zagros mountains, connecting Mesopotamia and the Iranian Plateau. The kingdoms of Aratta known from cuneiform sources may have been located in the Central Iranian Plateau.

In classical antquity the region was known as Persia, due to the Persian Achaemenid dynasty, originating in Persia proper, or Fars.

The Middle Persian *Erān* (whence Modern Persian *Irān*) began to be used in reference to the state (rather than as an ethnic designator) from the Sassanid period (see Etymology of Iran).

5.4 Archaeology

Further information: Prehistoric archaeological sites in Iran

Archaeological sites and cultures of the Iranian plateau include:

- Central Iranian Plateau ("Jiroft culture")

  - Shahr-i Sokhta

  - Konar Sandal

  - Tepe Yahya

- Zayandeh River Civilization

- Tappeh Sialk

- Paleolithic sites

  - Niasar
  - Sefid-Ab
  - Kaftar Khoun
  - Qaleh Bozi Caves
  - Mirak
  - Delazian
  - Tabas
  - Masileh

5.5   Flora

The plateau has historical oak and poplar forests. Oak forests are found around Shiraz. Aspen, elm, ash, willow, walnut, pine, and cypress are also found, though the latter two are rare. As of 1920, poplar was harvested for making doors. Elm was used for ploughs. Other trees like acacia, cypress, and Turkestan elm were used for decorative purposes. Flower wise, the plateau can grow lilac, jasmine, and roses. Hawthorn and Cercis siliquastrum are common, which are both used for basket weaving.*[7]

5.6   Economy

The Iranian plateau harvests trees for making doors, ploughs, and baskets. Fruit is grown also. Pears, apples, apricots, quince, plums, nectarines, cherries, mulberries, and peaches were commonly seen in the 20th century. Almonds and pistachios are common in warmer areas. Dates, oranges, grapes, melon, and limes are also grown. Other edibles include potatoes and cauliflower, which were hard to grow until European settlement brought irrigation improvements. Other vegetables include cabbage, tomatoes, artichokes, cucumbers, spinach, radishes, lettuce, and eggplants.*[7]

The plateau also produces wheat, barley, millet, beans, opium, cotton, lucerne, and tobacco. The barley is fed mainly to horses. Sesame is grown and made into sesame oil. Mushrooms and manna were also seen in the plateau area as of 1920. Caraway is grown in the Kerman Province.*[7]

5.7   See also

- Geography of Iran

- Greater Iran

- List of Iranian four-thousanders

5.8   References

[1] Robert H. Dyson. *The archaeological evidence of the second millennium B.C. on the Persian plateau.* ISBN 0-521-07098-8.

[2] James Bell (1832). *A System of Geography, Popular and Scientific.* Archibald Fullarton. pp. 7,284,287,288.

[3] Old Iranian Online, *University of Texas College of Liberal Arts* (retrieved 10 February 2007)

[4] s.v. "ancient Iran"

[5] s.v. "Elamite language"

[6] "Iranian Plateau". Peakbagger.com.

[7] Sykes, Percy (1921). *A History of Persia*. London: Macmillan and Company. pp. 75–76.

- Y. Majidzadeh, *Sialk III and the Pottery Sequence at Tepe Ghabristan. The Coherence of the Cultures of the Central Iranian Plateau,* Iran 19, 1981, 141-46.

5.9 External links

- "Iranian Plateau". Peakbagger.com.
- "Central Iranian Plateau". Peakbagger.com.

Chapter 6

Indus River

"Indus" and "Sindhu" redirect here. For other uses, see Indus (disambiguation) and Sindhu (disambiguation).
The **Indus River**, also called the **Sindhū River** (Sindhi: دريا سنڌو), or **Abāsīn** (Pashto: اباسين) in Khyber Pakhtunkhwa,

River Indus in Kharmang District, Pakistan.

is a major south-flowing river in South Asia. The total length of the river is 3,180 km (1,980 mi) which makes it one of longest rivers in Asia. It flows through Pakistan, the Indian state of Jammu and Kashmir and western Tibet. Originating in the Tibetan Plateau in the vicinity of Lake Mansarovar, the river runs a course through the Ladakh region of Jammu and Kashmir, towards Gilgit-Baltistan and then flows in a southerly direction along the entire length of Punjab, Pakistan to merge into the Arabian Sea near the port city of Karachi in Sindh. It is the longest river of Pakistan.

The river has a total drainage area exceeding 1,165,000 km^2 (450,000 sq mi). Its estimated annual flow stands at around 207 km^3 (50 cu mi), making it the twenty-first largest river in the world in terms of annual flow. The Zanskar is its left bank tributary in Ladakh. In the plains, its left bank tributary is the Chenab which itself has four major tributaries, namely, the Jhelum, the Ravi, the Beas and the Sutlej. Its principal right bank tributaries are the Shyok, the Gilgit, the Kabul, the Gomal and the Kurram. Beginning in a mountain spring and fed with glaciers and rivers in the Himalayas, the river supports ecosystems of temperate forests, plains and arid countryside.

The Indus forms the delta of present-day Pakistan mentioned in the Vedic Rigveda as *Sapta Sindhu* and the Iranian Zend Avesta as *Hapta Hindu* (both terms meaning "seven rivers"). The river has been a source of wonder since the Classical Period, with King Darius of Persia sending his Greek subject Scylax of Caryanda to explore the river as early as 510 BC.

6.1 Etymology and names

The word "Indus" is the romanised form of the ancient Greek word "Indós" (Ἰνδός), borrowed from the old Persian word "Hinduš"

Megasthenes's book *Indica* derives its name from the river's Greek name, "Indós" (Ἰνδός), and describes Nearchus's contemporaneous account of how Alexander the Great crossed the river. The ancient Greeks referred to the Indians (people of present-day India and Pakistan) as "Indói" (Ἰνδοί), literally meaning "the people of the Indus".[1] The country of India and the Pakistani province of Sindh owe their names to the river.[2]

6.1.1 Rigveda and the Indus

Rigveda also describes several mythical rivers, including one named "Sindhu". The Rigvedic "Sindhu" is thought to be the present-day Indus river and is attested 176 times in its text – 95 times in the plural, more often used in the generic meaning. In the Rigveda, notably in the later hymns, the meaning of the word is narrowed to refer to the Indus river in particular, as in the list of rivers mentioned in the hymn of *Nadistuti sukta*. The Rigvedic hymns apply a feminine gender to all the rivers mentioned therein but "Sindhu" is the only river attributed with a masculine gender. Sindhu is seen as a strong warrior amongst other rivers which are seen as goddesses and compared to cows and mares yielding milk and butter.

6.1.2 Other names

In Urdu, the official language of Pakistan, the Indus is known as دریائے سندھ (*Daryā-e Sindh*). In other languages of the region, the river is known as सिन्धु नदी (*Sindhu Nadī*) in Hindi, سنڌو (*Sindhu*) in Sindhi, سندہ (*Sindh*) in Shahmukhi alphabet, ਸਿੰਧ ਨਦੀ (*Sindh Nadī*) in Gurmukhī alphabet, સિંધુ નદી (*Sindhu*) in Gujarati; ابّاسين (*Abāsin*, lit. "Father of Rivers") in Pashto, رود سندھ, (*Nahar al-Sind*) in Arabic,(*Sêngê Zangbo*, lit. "Lion River") inTibetan,印度(*Yìndù*)inChinese,and*Nilab*inTurki.

6.2 Description

The Indus River provides key water resources for Pakistan's economy – especially the *breadbasket* of Punjab province, which accounts for most of the nation's agricultural production, and Sindh. The word Punjab means "land of five rivers" and the five rivers are Jhelum, Chenab, Ravi, Beas and Sutlej, all of which finally flow into the Indus. The Indus also supports many heavy industries and provides the main supply of potable water in Pakistan.

The ultimate source of the Indus is in Tibet; the river begins at the confluence of the Sengge and Gar rivers that drain the Nganglong Kangri and Gangdise Shan (Gang Rinpoche, Mt. Kailas) mountain ranges. The Indus then flows northwest through Ladakh and Baltistan into Gilgit, just south of the Karakoram range. The Shyok, Shigar and Gilgit rivers carry glacial waters into the main river. It gradually bends to the south, coming out of the hills between Peshawar and

Rawalpindi. The Indus passes gigantic gorges 4,500–5,200 metres (15,000–17,000 feet) deep near the Nanga Parbat massif. It flows swiftly across Hazara and is dammed at the Tarbela Reservoir. The Kabul River joins it near Attock. The remainder of its route to the sea is in the plains of the Punjab and Sindh, where the flow of the river becomes slow and highly braided. It is joined by the Panjnad at Mithankot. Beyond this confluence, the river, at one time, was named the **Satnad River** (*sat* = "seven" , *nadī* = "river"), as the river now carried the waters of the Kabul River, the Indus River and the five Punjab rivers. Passing by Jamshoro, it ends in a large delta to the east of Thatta.

The Indus is one of the few rivers in the world to exhibit a tidal bore. The Indus system is largely fed by the snows and glaciers of the Himalayas, Karakoram and the Hindu Kush ranges of Tibet, the Indian states of Jammu and Kashmir and Himachal Pradesh and Gilgit-Baltistan of Pakistan. The flow of the river is also determined by the seasons – it diminishes greatly in the winter, while flooding its banks in the monsoon months from July to September. There is also evidence of a steady shift in the course of the river since prehistoric times – it deviated westwards from flowing into the Rann of Kutch and adjoining Banni grasslands after the 1816 earthquake.[*][3][*][4]

The traditional source of the river is the *Senge Khabab* or "Lion's Mouth" , a perennial spring, not far from the sacred Mount Kailash marked by a long low line of Tibetan chortens. There are several other tributaries nearby, which may possibly form a longer stream than Senge Khabab, but unlike the Senger Khabab, are all dependent on snowmelt. The Zanskar River, which flows into the Indus in Ladakh, has a greater volume of water than the Indus itself before that point.[*][5]

> "That night in the tent [next to Senge Khabab] I ask Sonmatering which of the Indus tributaries which we crossed this morning is the longest. All of them, he says, start at least a day's walk away from here. The Bukhar begins near the village of Yagra. The Lamolasay's source is in a holy place: there is a monastery there. The Dorjungla is a very difficult and long walk, three days perhaps, and there are many sharp rocks; but it its water is clear and blue, hence the tributary's other name, Zom-chu, which Karma Lama translates as 'Blue Water'. The Rakmajang rises from a dark lake called the Black Sea.
>
> One of the longest tributaries —and thus a candidate for the river's technical source —is the Kla-chu, the river we crossed yesterday by bridge. Also known as the Lungdep Chu, it flows into the Indus from the south-east, and rises a day's walk from Darchen. But Sonamtering insists that the Dorjungla is the longest of the 'three types of water' that fall into the Seng Tsanplo ['Lion River' or Indus]." [*][5]

6.3   History

Main articles: Indus Valley Civilization and History of Sindh

Paleolithic sites have been discovered in Pothohar near Pakistan's capital Islamabad, with the stone tools of the Soan Culture. In ancient Gandhara, near Islamabad, evidence of cave dwellers dated 15,000 years ago has been discovered at Mardan.

The major cities of the Indus Valley Civilization, such as Harappa and Mohenjo-daro, date back to around 3300 BC, and represent some of the largest human habitations of the ancient world. The Indus Valley Civilization extended from across Pakistan and northwest India, with an upward reach from east of Jhelum River to Ropar on the upper Sutlej. The coastal settlements extended from Sutkagan Dor at the Pakistan, Iran border to Kutch in modern Gujarat, India. There is an Indus site on the Amu Darya at Shortughai in northern Afghanistan, and the Indus site Alamgirpur at the Hindon River is located only 28 km (17 mi) from Delhi. To date, over 1,052 cities and settlements have been found, mainly in the general region of the Ghaggar-Hakra River and its tributaries. Among the settlements were the major urban centers of Harappa and Mohenjo-daro, as well as Lothal, Dholavira, Ganeriwala, and Rakhigarhi. Only 90–96 of more than 800 known Indus Valley sites have been discovered on the Indus and its tributaries. The Sutlej, now a tributary of the Indus, in Harappan times flowed into the Ghaggar-Hakra River, in the watershed of which were more Harappan sites than along the Indus.

Most scholars believe that settlements of Gandhara grave culture of the early Indo-Aryans flourished in Gandhara from 1700 BC to 600 BC, when Mohenjo-daro and Harappa had already been abandoned.

The word "India" is derived from the Indus River. In ancient times, "India" initially referred to those regions immediately along the east bank of the Indus, but by 300 BC, Greek writers including Herodotus and Megasthenes were applying the term to the entire subcontinent that extends much farther eastward.[6][7]

The lower basin of the Indus forms a natural boundary between the Iranian Plateau and the Indian subcontinent; this region embraces all or parts of the Pakistani provinces Balochistan, Khyber Pakhtunkhwa, Punjab and Sindh and the countries Afghanistan and India. It was crossed by the invading armies of Alexander, but after his Macedonians conquered the west bank—joining it to the Hellenic Empire, they elected to retreat along the southern course of the river, ending Alexander's Asian campaign . The Indus plains were later dominated by the Persian empire and then the Kushan empire. Over several centuries Muslim armies of Muhammad bin Qasim, Mahmud of Ghazni, Mohammed Ghori, Tamerlane and Babur crossed the river to invade the inner regions of the Punjab and points farther south and east.

6.4 Geography

The Indus River near Leh, Ladakh, India

6.4.1 Tributaries

- Astor River
- Balram River
- Beas River
- Chenab River
- Dras River
- Gar River
- Ghizar River
- Gilgit River
- Gomal River
- Hunza River
- Jhelum River
- Kabul River
- Kurram River

- Nagar River

- Panjnad River

- Ravi River

- Satluj River

- Shigar River

- Shyok River

- Soan River

- Tanubal River

- Zanskar River

6.5   Geology

The Indus river feeds the Indus submarine fan, which is the second largest sediment body on the Earth at around 5 million cubic kilometres of material eroded from the mountains. Studies of the sediment in the modern river indicate that the Karakoram Mountains in northern Pakistan and India are the single most important source of material, with the Himalayas providing the next largest contribution, mostly via the large rivers of the Punjab (Jhelum, Ravi, Chenab, Beas and Sutlej). Analysis of sediments from the Arabian Sea has demonstrated that prior to five million years ago the Indus was not connected to these Punjab rivers which instead flowed east into the Ganges and were captured after that time.[8] Earlier work showed that sand and silt from western Tibet was reaching the Arabian Sea by 45 million years ago, implying the existence of an ancient Indus River by that time.[9] The delta of this proto-Indus river has subsequently been found in the Katawaz Basin, on the Afghan-Pakistan border.

In the Nanga Parbat region, the massive amounts of erosion due to the Indus river following the capture and rerouting through that area is thought to bring middle and lower crustal rocks to the surface.[10]

In November 2011, satellite images showed that Indus river has re-entered India feeding Great Rann of Kutch , Little Rann of Kutch and a lake near Ahmedabad known as Nal Sarovar.[11] Heavy rains had left the river basin along with the Lake Manchar, Lake Hemal and Kalri Lake (all in modern-day Pakistan) inundated. This incident happened after two centuries, when Indus river majorly shifted its course westwards after 1819 Rann of Kutch earthquake.

6.6   Wildlife

Accounts of the Indus valley from the times of Alexander's campaign indicate a healthy forest cover in the region, which has now considerably receded. The Mughal Emperor Babur writes of encountering rhinoceroses along its bank in his memoirs (the Baburnama). Extensive deforestation and human interference in the ecology of the Shivalik Hills has led to a marked deterioration in vegetation and growing conditions. The Indus valley regions are arid with poor vegetation. Agriculture is sustained largely due to irrigation works. Indus river and its watershed has a rich biodiversity. It is home to around 25 amphibian species and 147 species, 22 of which are only found in the Indus.[12]

6.6.1   Mammals

The blind Indus River Dolphin (*Platanista indicus minor*) is a sub-species of dolphin found only in the Indus River. It formerly also occurred in the tributaries of the Indus river. According to the World Wildlife Fund claims it is one of the most threatened cetaceans with only about 1000 still existing.[13]

6.6.2 Fish

Palla fish Tenualosa ilisha of the river is a delicacy for people living along the river. The population of fish in the river is moderately high, with Sukkur, Thatta and Kotri being the major fishing centres – all in the lower Sindh course. But damming and irrigation has made fish farming an important economic activity. Located southeast of Karachi, the large delta has been recognised by conservationists as one of the world's most important ecological regions. Here the river turns into many marshes, streams and creeks and meets the sea at shallow levels. Here marine fishes are found in abundance, including pomfret and prawns.

6.7   Economy

The Indus is the most important supplier of water resources to the Punjab and Sindh plains – it forms the backbone of agriculture and food production in Pakistan. The river is especially critical since rainfall is meager in the lower Indus valley. Irrigation canals were first built by the people of the **Indus valley civilization**, and later by the engineers of the Kushan Empire and the Mughal Empire. Modern irrigation was introduced by the British East India Company in 1850 – the construction of modern canals accompanied with the restoration of old canals. The British supervised the construction of one of the most complex irrigation networks in the world. The Guddu Barrage is 1,350 m (4,430 ft) long – irrigating Sukkur, Jacobabad, Larkana and Kalat. The Sukkur Barrage serves over 20,000 km^2 (7,700 sq mi).

After Pakistan came into existence, a water control treaty signed between India and Pakistan in 1960 guaranteed that Pakistan would receive water from the Indus River and its two tributaries the Jhelum River & the Chenab River independently of upstream control by India.[14]

The Indus Basin Project consisted primarily of the construction of two main dams, the Mangla Dam built on the Jhelum River and the Tarbela Dam constructed on the Indus River, together with their subsidiary dams.[15] The Pakistan Water and Power Development Authority undertook the construction of the Chashma-Jhelum link canal – linking the waters of the Indus and Jhelum rivers – extending water supplies to the regions of Bahawalpur and Multan. Pakistan constructed the Tarbela Dam near Rawalpindi – standing 2,743 metres (9,000 ft) long and 143 metres (470 ft) high, with an 80-kilometre (50 mi) long reservoir. The Kotri Barrage near Hyderabad is 915 metres (3,000 ft) long and provides additional supplies for Karachi. It support the Chashma barrage near Dera Ismail Khan use for irrigation and flood control. for The Taunsa Barrage near Dera Ghazi Khan produces 100,000 kilowatts of electricity. The extensive linking of tributaries with the Indus has helped spread water resources to the valley of Peshawar, in the Khyber Pakhtunkhwa. The extensive irrigation and dam projects provide the basis for Pakistan's large production of crops such as cotton, sugarcane and wheat. The dams also generate electricity for heavy industries and urban centres.

6.8   People

The inhabitants of the regions through which the Indus river passes and forms a major natural feature and resource are diverse in ethnicity, religion, national and linguistic backgrounds. On the northern course of the river in the state of Jammu and Kashmir in India, live the Buddhist people of Ladakh, of Tibetan stock, and the Dards of Indo-Aryan or Dardic stock and practising Buddhism and Islam. Then it descends into Baltistan, northern Pakistan passing the main Balti city of Skardu. On its course river from Dubair Bala also drains into it at Dubair Bazar. People living at this area are mainly Kohistani and speak Kohistani language. Major areas through which Indus river pass through in Kohistan are Dasu, Pattan and Dubair. As it continues through Pakistan, the Indus river forms a distinctive boundary of ethnicity and cultures – upon the western banks the population is largely Pashtun, Baloch, and of other Iranian stock. The eastern banks are largely populated by people of Indo-Aryan stock, such as the Punjabis and the Sindhis. In northern Punjab and the Khyber Pakhtunkhwa, ethnic Pashtun tribes live alongside Dardic people in the hills (Khowar, Kalash, Shina, etc.), Burushos (in Hunza), and Punjabi people.

Through its course in Punjab the people living along the Indus river are distinct from Punjabi and Pustoon. This distinction is not only based on language (Saraiki dialect) but also in their genealogy. They are tall and slender, distinctively different from either Pushtoon or Punjabi who have a sturdy built. These people live in Mianwali and Dera Ismail Khan, Dera Ghazi Khan, Rahim Yar Khan and Rajan Pur in Punjab. In the province of Sindh, upper third of River indus is again inhabited by Saraiki speaking people up to Shikapur. The rest of the Indus river valley is inhabited by Sindhis and Baloch of Sindhi language. Upon the western banks of the river live the Baloch and Pashtun people of Balochistan.

6.9   Modern issues

The Indus is a strategically vital resource for Pakistan's economy and society. After Pakistan and India declared Independence from the British Raj, also known as Indian Empire, the use of the waters of the Indus and its five eastern tributaries became a major dispute between India and Pakistan. The irrigation canals of the Sutlej valley and the Bari Doab were split – with the canals lying primarily in Pakistan and the headwork dams in India disrupting supply in some parts of Pakistan. The concern over India building large dams over various Punjab rivers that could undercut the supply flowing to Pakistan, as well as the possibility that India could divert rivers in the time of war, caused political consternation in Pakistan. Holding diplomatic talks brokered by the World Bank, India and Pakistan signed the Indus Waters Treaty in 1960. The treaty gave India control of the three easternmost rivers of the Punjab, the Sutlej, the Beas and the Ravi, while Pakistan gained control of the three western rivers, the Jhelum, the Chenab and the Indus. India retained the right to use of the western rivers for non-irrigation projects. (See discussion regarding a recent dispute about a hydroelectric project on the Chenab (not Indus) known as the Baglihar Project).

There are concerns that extensive deforestation, industrial pollution and global warming are affecting the vegetation and wildlife of the Indus delta, while affecting agricultural production as well. There are also concerns that the Indus river may be shifting its course westwards – although the progression spans centuries. On numerous occasions, sediment clogging owing to poor maintenance of canals has affected agricultural production and vegetation. In addition, extreme heat has caused water to evaporate, leaving salt deposits that render lands useless for cultivation.

Recently, India's construction of dams on the river, which Pakistan claims is in violation of the Indus Waters Treaty reducing water flow into Pakistan, has caused Pakistan to take the issue to the international courts for arbitration.

6.9.1   Effects of climate change on the river

The Tibetan Plateau contains the world's third-largest store of ice. Qin Dahe, the former head of the China Meteorological Administration, said the recent fast pace of melting and warmer temperatures will be good for agriculture and tourism in the short term, but issued a strong warning:

> "Temperatures are rising four times faster than elsewhere in China, and the Tibetan glaciers are retreating at a higher speed than in any other part of the world... In the short term, this will cause lakes to expand and bring floods and mudflows.. In the long run, the glaciers are vital lifelines of the Indus River. Once they vanish, water supplies in Pakistan will be in peril." *[16]

"There is insufficient data to say what will happen to the Indus," says David Grey, the World Bank's senior water advisor in South Asia. "But we all have very nasty fears that the flows of the Indus could be severely, severely affected by glacier melt as a consequence of climate change," and reduced by perhaps as much as 50 percent. "Now what does that mean to a population that lives in a desert [where], without the river, there would be no life? I don't know the answer to that question," he says. "But we need to be concerned about that. Deeply, deeply concerned." *[17]

6.9.2   Pollution

Over the years factories on the banks of the Indus River have increased levels of water pollution in the river and the atmosphere around it. High levels of pollutants in the river have led to the deaths of endangered Indus River Dolphin. The Sindh Environmental Protection Agency has ordered polluting factories around the river to shut down under the Pakistan Environmental Protection Act, 1997.*[18] Death of the Indus River Dolphin has also been attributed to fishermen using poison to kill fish and scooping them up.*[19]*[20] As a result, the government banned fishing from Guddu Barrage to Sukkur.*[21]

6.9.3   2010 floods

Main article: 2010 Pakistan floods

In July 2010, following abnormally heavy monsoon rains, the Indus River rose above its banks and started flooding. The rain continued for the next two months, devastating large areas of Pakistan. In Sindh, the Indus burst its banks near Sukkur on 8 August, submerging the village of Mor Khan Jatoi.[22] In early August, the heaviest flooding moved southward along the Indus River from severely affected northern regions toward western Punjab, where at least 1,400,000 acres (570,000 ha) of cropland was destroyed, and the southern province of Sindh.[23] As of September 2010, over two thousand people had died and over a million homes had been destroyed since the flooding began.[24][25]

6.9.4   2011 floods

Main article: 2011 Sindh floods

The 2011 Sindh floods began during the Pakistani monsoon season in mid-August 2011, resulting from heavy monsoon rains in Sindh, eastern Balochistan, and southern Punjab.[26] The floods caused considerable damage; an estimated 434 civilians were killed, with 5.3 million people and 1,524,773 homes affected.[27] Sindh is a fertile region and often called the "breadbasket" of the country; the damage and toll of the floods on the local agrarian economy was said to be extensive. At least 1.7 million acres (690,000 ha; 2,700 sq mi) of arable land were inundated. The flooding followed the previous year's floods, which devastated a large part of the country.[27] Unprecedented torrential monsoon rains caused severe flooding in 16 districts of Sindh.[28]

6.10   See also

- 1974 Hunza earthquake
- Chura Sharif
- Ghaggar-Hakra River
- HMS Indus, ships named after the Indus River
- Indus Valley Civilization
- Indus Waters Treaty
- Sindhology
- Sindhu Darshan Festival
- Sarasvati River
- Sindhu Pushkaram
- Sind Sagar Doab
- Rigvedic rivers

6.11   Notes

[1] Kuiper 2010, p. 86.

[2] Encyclopædia Britannica.

[3] 70% of cattle-breeders desert Banni; by Narandas Thacker, TNN, 14 February 2002; The Times of India

[4] Lost and forgotten: grasslands and pastoralists of Gujarat; by Charul Bharwada and Vinay Mahajan; The forsaken drylands; a symposium on some of India's most invisible people; Seminar; New Delhi; 2006; NUMB 564, pages 35–39; ISSN 0037-1947, Listed at the British Library Online

[5] Albinia (2008), p. 307.

[6] Henry Yule: *India, Indies. In Hobson-Jobson: A glossary of colloquial Anglo-Indian words and phrases, and of kindred terms, etymological, historical, geographical and discursive.* New ed. edited by William Crooke, B.A. London: J. Murray, 1903

[7] "Was the Ramayana actually set in and around today' s Afghanistan?".

[8] Clift, Peter D.; Blusztajn, Jerzy (15 December 2005). "Reorganization of the western Himalayan river system after five million years ago". *Nature* **438** (7070): 1001–1003. doi:10.1038/nature04379. PMID 16355221.

[9] Clift, Peter D.; Shimizu, N.; Layne, G.D.; Blusztajn, J.S.; Gaedicke, C.; Schlüter, H.-U.; Clark, M.K.; Amjad, S. (August 2001). "Development of the Indus Fan and its significance for the erosional history of the Western Himalaya and Karakoram". *GSA Bulletin* **113** (8): 1039–1051. doi:10.1130/0016-7606(2001)113<1039:DOTIFA>2.0.CO;2.

[10] Zeitler, Peter K.; Koons, Peter O.; Bishop, Michael P.; Chamberlain, C. Page; Craw, David; Edwards, Michael A.; Hamidullah, Syed; Jam, Qasim M.; Kahn, M. Asif; Khattak, M. Umar Khan; Kidd, William S. F.; Mackie, Randall L.; Meltzer, Anne S.; Park, Stephen K.; Pecher, Arnaud; Poage, Michael A.; Sarker, Golam; Schneider, David A.; Seeber, Leonardo; and Shroder, John F. (October 2001). "Crustal reworking at Nanga Parbat, Pakistan: Metamorphic consequences of thermal-mechanical coupling facilitated by erosion". *Tectonics* **20** (5): 712–728. doi:10.1029/2000TC001243.

[11] "Indus re-enters India after two centuries, feeds Little Rann, Nal Sarovar". India Today. 7 November 2011. Retrieved 2011-11-07.

[12] "Indus River" (PDF). *World' top 10 rivers at risk.* WWF. Retrieved 11 July 2012.

[13] "WWF – Indus River Dolphin". Wwf.panda.org. Retrieved 2012-09-22.

[14] "Tarabela Dam". www.structurae.the cat in the hat. Retrieved 2007-07-09.

[15] "Indus Basin Project". Encyclopædia Britannica. Retrieved 2007-07-09.

[16] "Global warming benefits to Tibet: Chinese official. Reported 18 August 2009". Google.com. 17 August 2009. Retrieved 2012-12-04.

[17] Pulitzercenter.org

[18] "SEPA orders polluting factory to stop production". *Dawn.* 3 Dec 2008. Retrieved 28 June 2012.

[19] "Fishing poison killing Indus dolphins, PA told". *Dawn.* 3/9/2012. Retrieved 28 June 2012. Check date values in: |date= (help)

[20] "'18 dolphins died from poisoning in Jan'". *Dawn.* 1 May 2012. Retrieved 28 June 2012.

[21] "Threat to dolphin: Govt bans fishing between Guddu and Sukkur". *The Express Tribune.* 9 Mar 2012. Retrieved 28 June 2012.

[22] Bodeen, Christopher (8 August 2010). "Asia flooding plunges millions into misery". Associated Press. Retrieved 8 August 2010.

[23] Guerin, Orla (7 August 2010). "Pakistan issues flooding 'red alert' for Sindh province". British Broadcasting Corporation. Retrieved 7 August 2010.

[24] "BBC News – Pakistan floods: World Bank to lend $900m for recovery". bbc.co.uk. 17 August 2010. Retrieved 2010-08-24.

[25] "BBC News – Millions of Pakistan children at risk of flood diseases". bbc.co.uk. 16 August 2010. Retrieved 2010-08-24.

[26] "Pakistan floods: Oxfam launches emergency aid response". *BBC World News South Asia.* 14 September 2011. Retrieved 15 September 2011.

[27] "Floods worsen, 270 killed: officials". *The Express Tribune.* 13 September 2011. Retrieved 13 September 2011.

[28] Government of Pakistan Pakmet.com.pk Retrieved on 19 September 2011 Archived April 24, 2012 at the Wayback Machine

6.12 References

- Albinia, Alice. (2008) *Empires of the Indus: The Story of a River*. First American Edition (20101) W. W. Norton & Company, New York. ISBN 978-0-393-33860-7.

- This article incorporates text from a publication now in the public domain: Chisholm, Hugh, ed. (1911). *Encyclopædia Britannica* (11th ed.). Cambridge University Press.

- World Atlas, Millennium Edition, pg 265

- Jean Fairley, "The Lion River", Karachi, 1978

6.13 External links

- Blankonthemap The Northern Kashmir WebSite

- Bibliography on Water Resources and International Law Peace Palace Library

- Northern Areas Development Gateway

- The Mountain Areas Conservancy Project

- Indus River watershed map (World Resources Institute)

- Indus Treaty

- Baglihar Dam issue

- Indus

- Indus Wildlife at the Wayback Machine (archived October 7, 2006)

- First raft and kayak descents of the Indus headwaters in Tibet

- Pulitzer Center on Crisis Reporting's project on water issues in South Asia

Babur crossing the Indus River.

Indus Valley archaeological sites

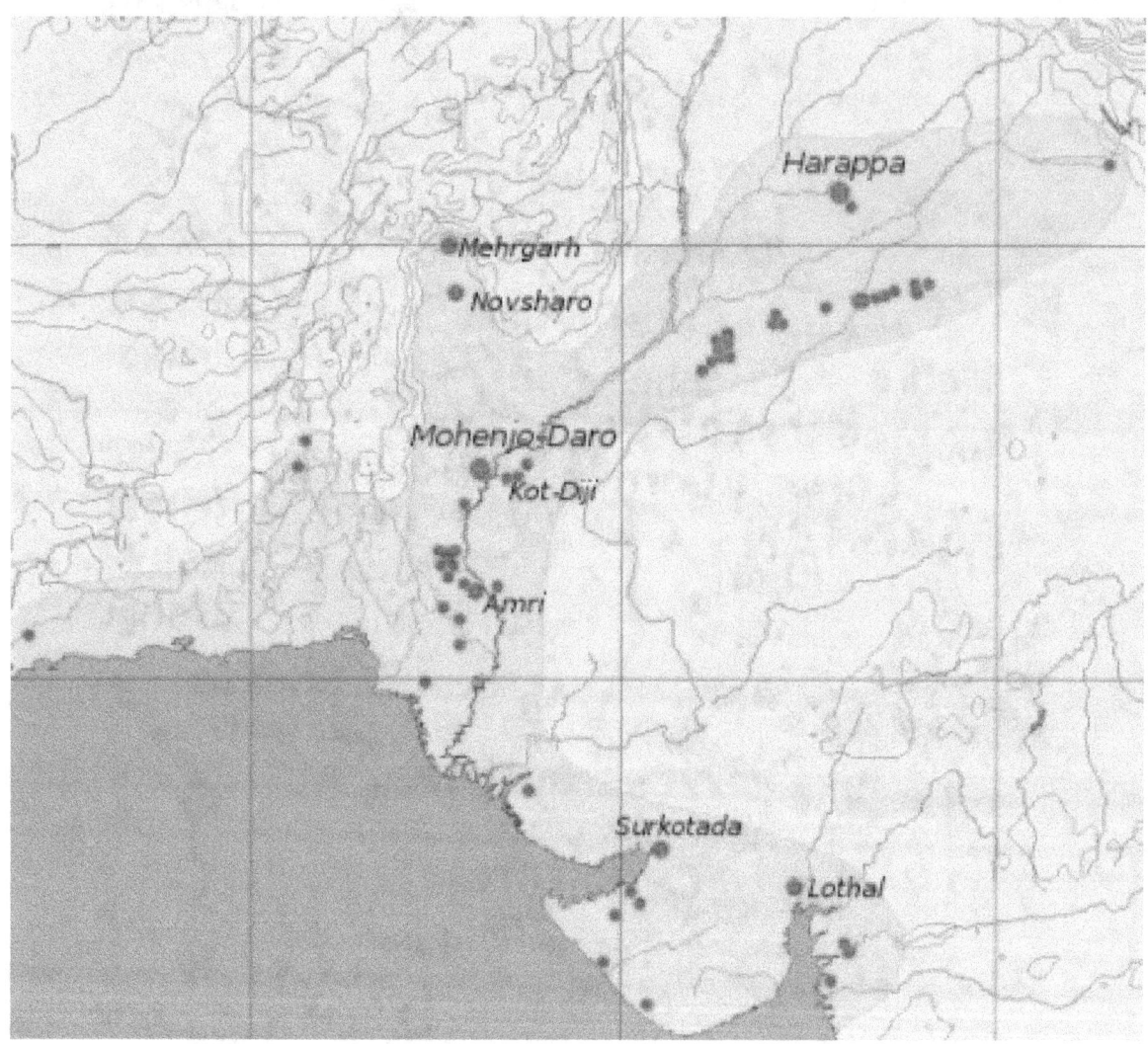

Extent and major sites of the Indus Valley Civilization 3000 BC

River Indus viewed from the Karakoram Highway.

Indus river near Leh, India, 2014

Confluence of Indus and Zanskar rivers. The Indus is at the bottom of the picture, flowing left-to-right; the Zanskar, carrying more water, comes in from the middle left of the picture.

Footbridge on the Indus River in Pakistan

Fishermen on the Indus River, c. 1905

The Indus River near Skardu, in Gilgit–Baltistan.

The Dubair Khwarr, a tributary of the Indus, near Shaikhdara, in Khyber Pakhtunkhwa.

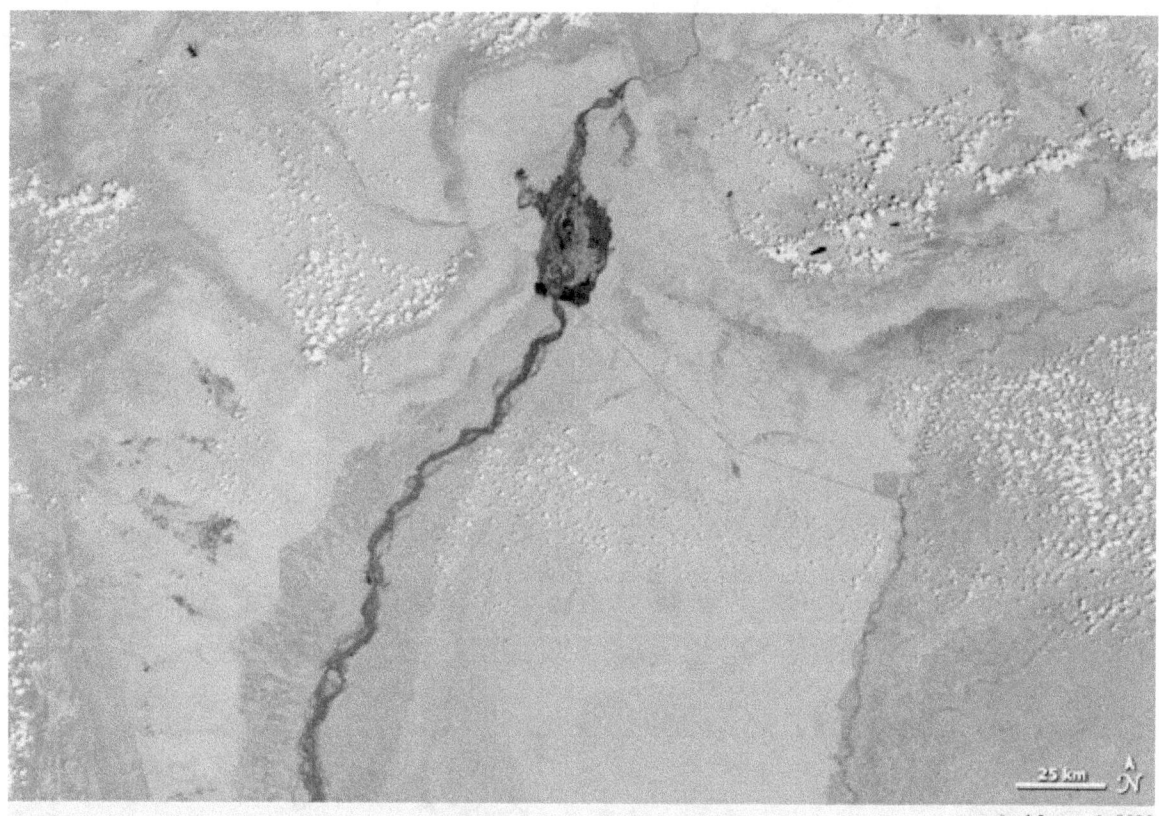

acquired August 1, 2009

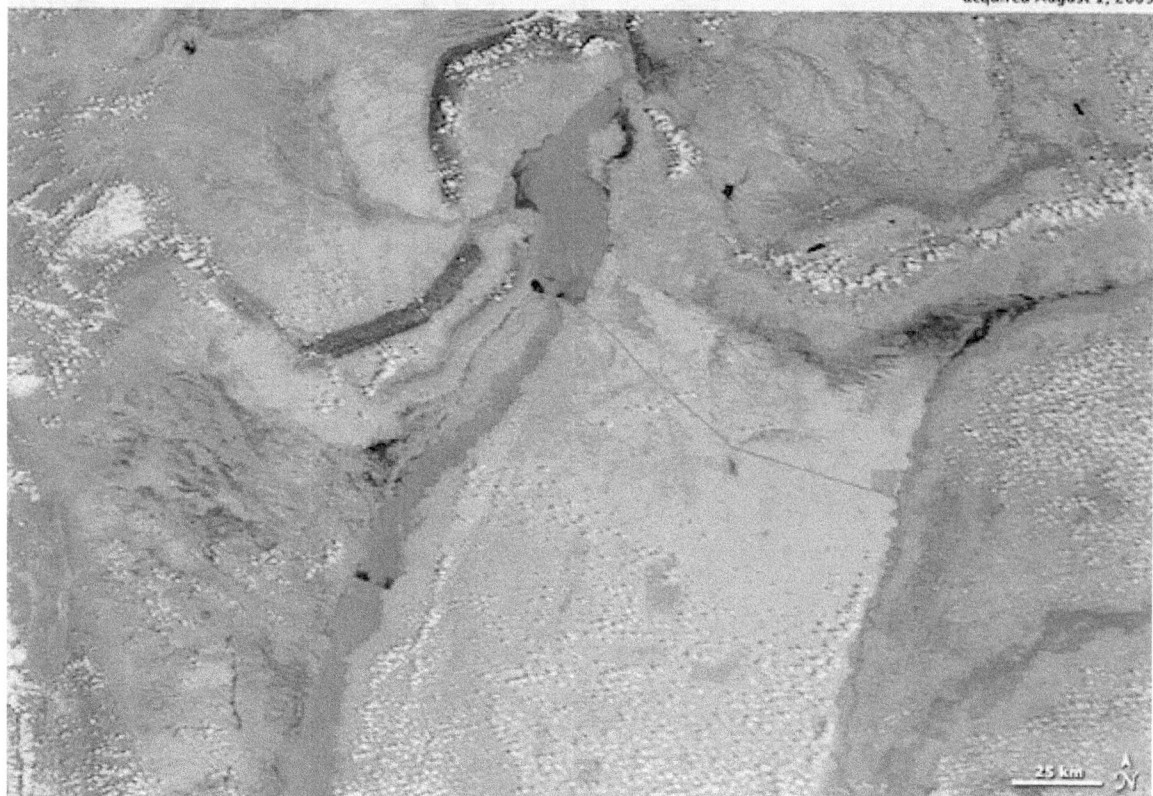

acquired July 31, 2010

Satellite images of the upper Indus River valley, comparing water-levels on 1 August 2009 (top) and 31 July 2010 (bottom)

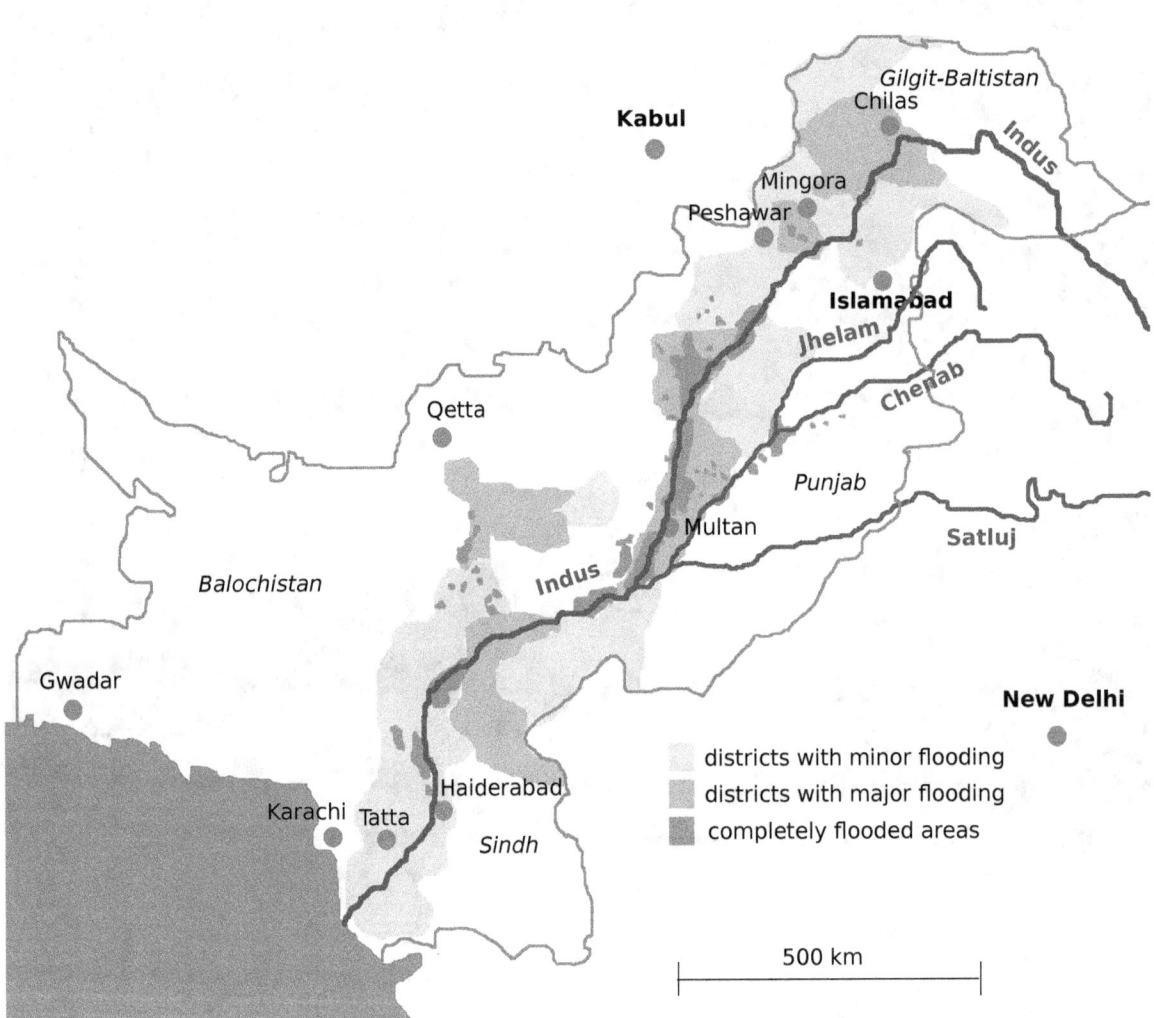

Affected areas as of 26 August 2010

Chapter 7

Prehistory of Anatolia

The **prehistory of Anatolia** stretches from 1.2 million years ago[1] through to the appearance of classical civilisation in the middle of the 1st millennium BC. It is generally regarded as being divided into three ages reflecting the dominant materials used for the making of domestic implements and weapons: Stone Age, Bronze Age and Iron Age. The term Copper Age (Chalcolithic) is used to denote the period straddling the stone and Bronze Ages.

Anatolia (*Turkish: Anadolu*), known by the Latin name of *Asia Minor*, is considered to be the westernmost extent of Western Asia. Geographically it encompasses the central uplands of modern Turkey, from the coastal plain of the Aegean Sea east to the mountains on the Armenian border and from the narrow coast of the Black Sea south to the Taurus mountains and Mediterranean coast.

The earliest representations of culture in Anatolia can be found in several archaeological sites located in the central and eastern part of the region. Stone Age artifacts such as animal bones and food fossils were found at Burdur (north of Antalya). Although the origins of some of the earliest peoples are shrouded in mystery, the remnants of Bronze Age civilizations such as the Hattian, Akkadian, Assyrian, and Hittite peoples provide us with many examples of the daily lives of its citizens and their trade. After the fall of the Hittites, the new states of Phrygia and Lydia stood strong on the western coast as Greek civilization began to flourish. Only the threat from a distant Persian kingdom prevented them from advancing past their peak of success.

7.1 Stone Age

The Stone Age is a prehistoric period in which stone was widely used in the manufacture of implements. This period occurred after the appearance of the genus *Homo* about 2.6 million years ago and roughly lasted 2.5 million years to the period between 4,500 and 2,000 BCE with the appearance of metalworking.

7.1.1 Paleolithic

In 2014, a stone tool was found in the Gediz River that was securely dated to 1.2 million years ago.[1] Evidence of paleolithic (prehistory 500,000 - 10,000 BCE) habitation include the Yarimburgaz Cave (Istanbul), Karain Cave (Antalya), and the Okuzini, Beldibi and Belbasi, Kumbucagi and Kadiini caves in adjacent areas. Examples of paleolithic humans can be found in the Museum of Anatolian Civilizations (Ankara), in the Archaeological Museum in Antalya, and in other Turkish institutions.

Evidence of fruit and of animal bones has been found at Yarimburgaz. The caves of the Mediterranean region contain murals.[2] Original claims (1975) of 250,000-year-old, Middle Pleistocene, *Homo sapiens* footprints at Kula[3] and Karain Caves are now considered erroneous and have been revised to the Late Pleistocene era.[4]

7.1.2 Mesolithic

Remains of a mesolithic culture in Anatolia can be found along the Mediterranean coast and also in Thrace and the western Black Sea area. Mesolithic remains have been located in the same caves as the paleolithic artefacts and drawings. Additional findings come from the Sarklimagara cave in Gaziantep, the Baradiz cave (Burdur), as well as the cemeteries and open air settlements at Sogut Tarlasi, Biris (Bozova) and Urfa.[*][5]

7.1.3 Neolithic

Further information: Anatolian hypothesis

Because of its strategic location at the intersection of Asia and Europe, Anatolia has been the center of several civilizations since prehistoric times. Neolithic settlements include Çatalhöyük, Çayönü, Nevali Cori, Aşıklı Höyük, Boncuklu Höyük Hacilar, Göbekli Tepe, Norsuntepe, Kosk, and Mersin.

Çatalhöyük (Central Turkey) is considered the most advanced of these, and Çayönü in the east the oldest (c. 7250 - 6750 BCE). We have a good idea of the town layout at Çayönü, based on a central square with buildings constructed of stone and mud. Archeological finds include farming tools that suggest both crops and animal husbandry as well as domestication of the dog. Religion is represented by figurines of Cybele, a mother goddess. Hacilar (Western Turkey) followed Çayönü, and has been dated to 7040 BCE.[*][6]

7.2 Chalcolithic (Copper) Age

Straddling the Neolithic and early Bronze Age, the Chalcolithic era (c. 5500 - 3000 BCE) is defined by the first metal implements made with copper. This age is represented in Anatolia by sites at Hacilar, Beycesultan, Canhasan, Mersin Yumuktepe, Elazig Tepecik, Malatya Degirmentepe, Norsuntepe, and Istanbul Fikirtepe.[*][7]

7.3 Bronze Age

Main article: Hattians
See also: List of Hattian and Hittite Kings

The Bronze Age (c. 3300 - 1200 BC) is characterised by the use of copper and its tin alloy, bronze, for manufacturing implements. Asia Minor was one of the first areas to develop bronze making.

7.3.1 Early Bronze Age

3000 - 2500 BCE Although the first habitation appears to have occurred as early as the 6th millennium BC during the Chalcolithic period, functioning settlements trading with each other occurred during the 3rd millennium BC. A settlement on a high ridge would become known as Büyükkaya, and later as the city of Hattush, the center of this civilization. Later, still, it would become the Hittite stronghold of Hattusha and is now Boğazköy. Remnants of Hattian civilization have been found both under the lower city of Hattusha and in the higher areas of Büyükkaya and Büyükkale,[*][8] Another settlement was established at Yarikkaya, about 2 km to the northeast.

The discovery of mineral deposits in this part of Anatolia allowed the Anatolians to develop metallurgy, such as the implements found in the royal graves at Alaca Höyük, about 25 km from Boğazköy, which it preceded, dating from 2400-2200 BC. Other Hattian centers include Hassum, Kanesh, Purushanda, and Zalwar.[*][9][*][10][*][11][*][12][*][13] During this time the Hattians engaged in trade with city states such as those of Sumer, which needed timber products from the Amanus mountains.

A royal tomb in Alaca Höyük

Anatolia had remained in the prehistoric period until it entered the sphere of influence of the Akkadian Empire in the 24th century BC under Sargon I, particularly in eastern Anatolia. However the Akkadian Empire suffered problematic climate changes in Mesopotamia, as well as a reduction in available manpower that affected trade. This led to its fall around 2150 BCE at the hands of the Gutians.*[14] The interest of the Akkadians in the region as far as it is known was for exporting various materials for manufacturing. Bronze metallurgy had spread to Anatolia from the Transcaucasian Kura-Araxes culture in the late 4th millennium BCE.*[15] While Anatolia was well endowed with copper ores, there was no evidence of substantial workings of the tin required to make bronze in Bronze-Age Anatolia.*[16]

7.3.2 Middle Bronze Age

2500 - 2000 BCE

At the origins of written history, the Anatolian plains inside the area ringed by the Kızılırmak River were occupied by the first defined civilization in Anatolia, a non-Indo-European indigenous people named the Hattians (c. 2500 BC – c. 2000 BC). During the middle Bronze Age, the Hattian civilization, including its capital of Hattush, continued to expand.*[10] The Anatolian middle Bronze Age influenced the early Minoan culture of Crete (3400 to 2200 BC) as evidenced by archaeological findings at Knossos.*[17]

7.3.3 Late Bronze Age

2000 - 1200 BCE

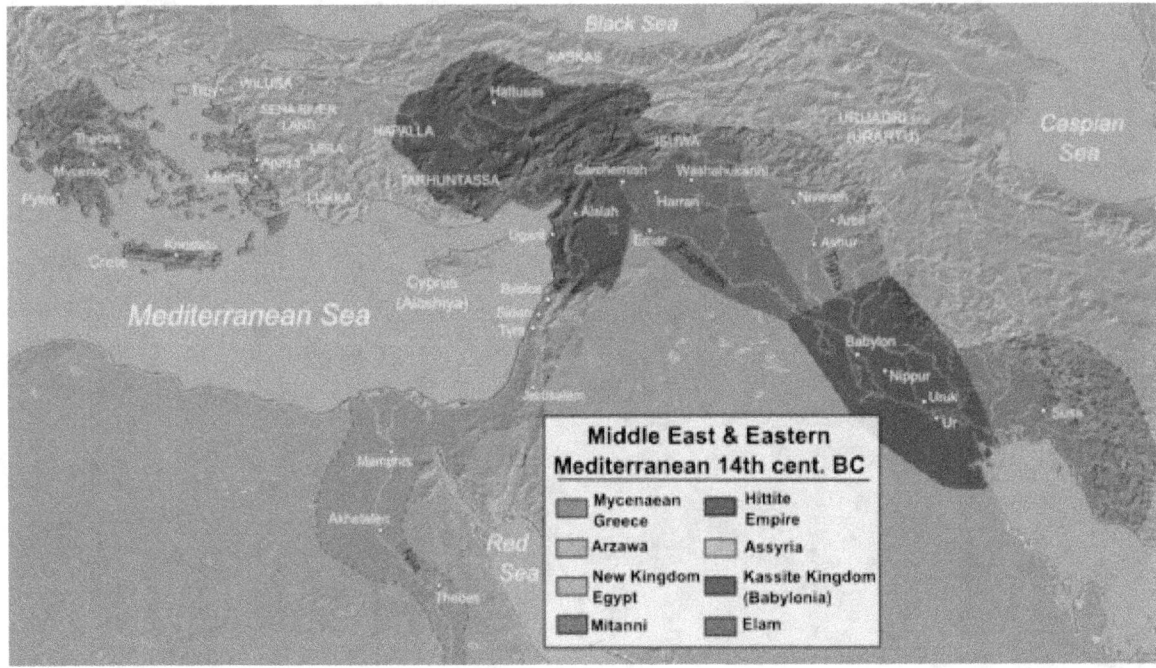

Map of the Ancient Near East during the Amarna Period (14th century BC), showing the great powers of the day: Egypt (yellow), Hatti (blue), the Kassite kingdom of Babylon (black), Assyria (yellow), and Mitanni (brown). The extent of the Achaean/Mycenaean civilization is shown in purple and of Arzawa in light green.

Hattians

The Hattians came into contact with Assyrians traders from Assur in Mesopotamia such as at Kanesh (Nesha) near modern Kültepe who provided them with the tin needed to make bronze. These trading posts or *Karums* (Akkadian for *Port*), have lent their name to a period, the Karum Period. The *Karums*, or Assyrian trading colonies, persisted in Anatolia until Hammurabi conquered Assyria and it fell under Babylonian domination in 1756 BC. These Karums represented separate residential areas where the traders lived, protected by the Hattites, and paying taxes in return. Meanwhile the fortifications of Hattush were strengthened with construction of royal residences on Büyükkale.

After the Assyrians overthrew their Gutians neighbours (c. 2050 BC) they claimed the local resources, notably silver, for themselves. However the Assyrians brought writing to Anatolia, a necessary tool for trading and business. These transactions were recorded in Akkadian cuneiform on clay tablets. Records found at Kanesh use an advanced system of trading computations and credit lines. The records also indicate the names of the cities where the transaction occurred.[*][15]

Hittites

Main article: History of the Hittites
Further information: Kizzuwatna and Arzawa
 The history of the Hittite civilization is known mostly from cuneiform texts found in the area of their empire, and from diplomatic and commercial correspondence found in various archives in Egypt and the Middle East.

Old Kingdom   Hattian civilization was also impacted by an invading Indo-European people, the Hittites, in the early 18th century BC, Hattush being burned to the ground in 1700 BC by King Anitta of Kussar after overthrowing King Piyushti. He then placed a curse on the site and set up his capital at Kanesh 160 km south east.[*][10] The Hittites absorbed the Hattians over the next century, a process that was essentially complete by 1650 BC. Eventually Hattusha became a Hittite centre by the second half of the 17th century BC, and King Hattusili I (1586–1556 BC) moved his capital back to Hattusha from Neša (Kanesh).

A drawing of rock-carved reliefs of a procession of Hittite deities in Yazılıkaya, Turkey.

The Old Hittite Empire (17th - 15th centuries BCE) was at its height in the 16th century BCE, encompassing central Anatolia, north-western Syria as far as Ugarit, and upper Mesopotamia. Kizzuwatna in southern Anatolia controlled the region separating the Hittite Empire from Syria, thereby greatly affecting trade routes. The peace was kept in accordance with both empires through treaties that established boundaries of control.

Middle Kingdom   Following the reign of Telipinu (c. 1460 BC) the Hittite kingdom entered a relatively weak and poorly documented phase, known as the Middle Kingdom, from the reign of Telipinu's son-in-law, Alluwamna (mid-15th century BC) to that of Muwatalli I (c. 1400 BC).

New Kingdom   King Tudhaliya I (early 14th century BC) ushered in a new era of Hittite power, often referred to as the Hittite Empire. The Kings took on a divine role in Hittite society and the Hittite peoples, often allied with neighbours such as the Kizzuwatna began to expand again, moving into Western Anatolia, absorbing the Luwian state of Arzawa and the Assuwa League.

It was not until the reign of King Suppiluliumas (c. 1344 – 1322 BC) that Kizzuwatna was taken over fully, although the Hittites still preserved their cultural accomplishments in Kummanni (now Şar, Turkey) and Lazawantiya, north of Cilicia.*[18]

In the 13th century, after the reign of Hattusili III (c. 1267 – 1237 BC), Hittite power began to wain, threatened by Egypt to the South and Assyria to the East, effectively ending with Suppiluliuma II (c. 1207 – 1178 BC).

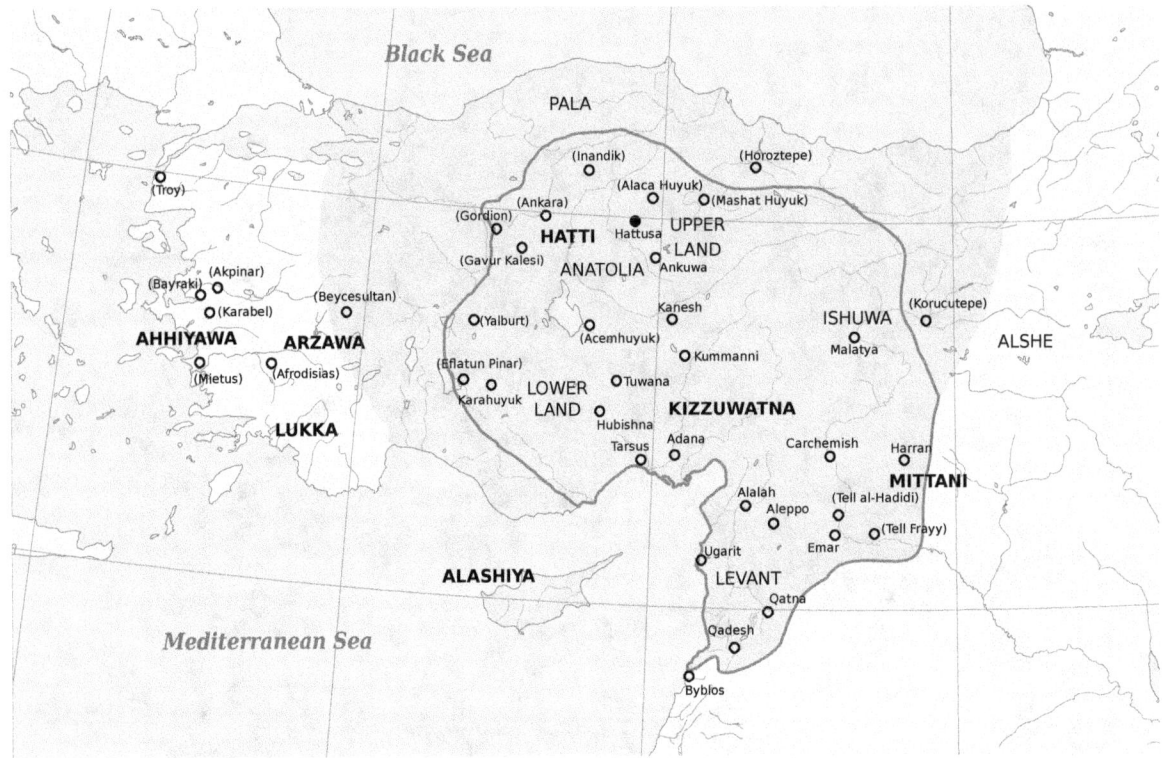

Approximate extent of Hittite rule, c. 1350-1300 BC, with Arzawa rule, Lukkans, Ahhiyawa to the west, Mitanni rule to the South-East.

Syro-Hittite era   Main articles: History of the Hittites and Bronze Age collapse

After 1180s BCE, amid general turmoil in the Levant associated with the sudden arrival of the Sea Peoples, and the collapse of the Bronze Age the empire disintegrated into several independent Syro-Hittite (Neo-Hittite) city-states, some of which survived until as late as the 8th century BCE. In the West, Greeks were arriving on the Anatolian coast, and the Kaskas along the northern Black Sea coast. Eventually Hattusha itself was destroyed around 1200 BCE and the age of Empires shifted to that of regional states as the Bronze Age transitioned into the Iron Age.

7.4 Iron Age

The Iron Age (c. 1300 – 600 BC) was characterised by the widespread use of iron and steel. It is also an age known for the development of alphabets and early literature. It formed the last phase of Pre-history, spanning the period between the collapse of the Bronze Age and the rise of classical civilisation. In Anatolia the dissolution of the Hittite Empire was replaced by regional Neo-Hittite powers, including Troad, Ionia, Lydia, Caria and Lycia in the west, Phrygia, centrally and Cimmeria and Urartu in the north east, while the Assyrians occupied much of the south east.

7.4.1 Western Anatolia

See also: Mysia and Doris (Asia Minor)

Regions of Anatolia, c. 500 BC. Aegean Greek settlements italicised

Troad

Main articles: Troad and Troy

The Troad, on the Biga peninsula, was the northernmost of the Aegean settlements in this period, best known for the legendary and historical city state of Troy. There were probably settlements in this region dating back to 3000 BC and the various archeological layers representing successive civilisations are referred to as Troy I (3000–2600 BC) to Troy IX (1st Century BC). Iron Age Troy corresponds to Troy VII-VIII, and coincides with the Homeric account of Troy and the Trojan Wars.

Aeolis

Main article: Aeolis

Aeolis was an area of the north western Aegean coast, between Troad and Ionia, from the Hellespont to the Hermus River (Gediz), west of Mysia and Lydia. By the 8th century BC the twelve most important cities formed a league. In the 6th century the cities were progressively conquered by Lydia, and then Persia.

Ionia

Main article: Ionia

Ionia was part of a group of settlements on the central Aegean coast bounded by Lydia to the east, and Caria to the south, known as the Ionian league. Ionians had been expelled from the Peloponnesus by the Dorians, and were resettled on the Aegean coastline of Anatolia by the Athenians to whose land they had fled. By the time of the last Lydian king, Croesus

(560–545 BC, Ionia fell under Lydian, and then Persian rule. With the defeat of Persia by the Greeks, Ionia again became independent until absorbed into the Roman province of Asia.

Lydia (Maeonia)

Main article: Lydia
Further information: List of kings of Lydia
 Lydia, or Maeonia as it was called before 687 BCE, was a major part of the history of western Anatolia, beginning with the Atyad dynasty, who first appeared around 1300 BCE. Lydia was situated to the west of Phrygia and east of the Aegean settlement of Ionia. The Lydians were Indo-European, speaking an Anatolian language related to Luwian and Hittite.

The Heraclids, managed to rule successively from 1185-687 BCE despite a growing presence of Greek influences along the Mediterranean coast. As Greek cities such as Smyrna, Colophon, and Ephesus rose, the Heraclids became weaker and weaker. The last king, Candaules, was murdered by his friend and lance-bearer named Gyges, and he took over as ruler. Gyges waged war against the intruding Greeks, and soon faced by a grave problem as the Cimmerians began to pillage outlying cities within the kingdom. It was this wave of attacks that led to the incorporation of the formerly independent Phrygia and its capital Gordium into the Lydian domain. It was not until the successive rules of Sadyattes and Alyattes II, ending in 560 BCE, that the attacks of the Cimmerians ended for good.

Under the reign of the last Lydian king Croesus, Lydia reached its greatest expansion. Persia was invaded first at the Battle of Pteria ending without a victor. Progressing deeper into Persia, Croesus was thoroughly defeated in the Battle of Thymbra at the hands of the Persian Cyrus II in 546 BC.[*][19]

Following Croesus' defeat, Lydia fell under the hegemony of Persia, Greece, Rome and Byzantium until finally being absorbed into the Turkish lands.

Caria

Main article: Caria

Caria forms a region in Western Anatolia, south of Lydia, east of Ionia and north of Lycia. Partially Greek (Ionian and Dorian), and possibly partially Minoan. Caria became subject to Persia, Greece and Rome before being absorbed into Byzantium. Remnants of the Carian civilisation form a rich legacy in the south western Aegean. Caria managed to maintain a relative degree of independence during successive occupation, and its symbol, the double headed axe is seen as a mark of defiance and can be seen inscribed on many buildings. The mausoleum at Halicarnassus (modern Bodrum), the tomb of the Persian Satrap Mausolus, was considered one of the Seven Wonders of the Ancient World. Other important relics include that of Mylasa (Milas) at one time capital of Caria and administrative seat of Mausolus, Labranda in the mountains high above Mylasa and Euromos (Herakleia) near Lake Bafa.

Lycia

Main article: Lycia

Lycia formed the southernmost settlement in Western Anatolia on what is now the Teke peninsula on the western Mediterranean coast. There many historic Lycian sites include Xanthos, Patara, Myra, Pinara, Tlos, Olympos and Phaselis. Emerging at the end of the Bronze Age as a Neo-Hittite league of city states whose governance model still influences political systems today. Alternating between Persian and Greek rule it eventually was incorporated into Rome, Byzantium and the Turkish lands.

7.4.2 Central Anatolia

See also: Bithynia, Paphlagonia, Galatia, Cappadocia, Lycaonia, Pisidia, Pamphylia, Pontus and Cilicia

Phrygia

Main article: Phrygia

 The west-central area of Anatolia became the domain of the Phrygian Kingdom following the fragmentation of the Hittite Empire during the 12th century BCE, existing independently until the 7th century BCE, and strongly featured in Greek mythology. Although their origin is disputed, their language more resembled Greek (Dorian) than the Hittites whom they succeeded. Possibly from the region of Thrace, the Phrygians eventually established their capital at Gordium (now Yazılıkaya). Known as Mushki by the Assyrians, the Phrygian people lacked central control in their style of government, and yet established an extensive network of roads. They also held tightly onto a lot of the Hittite facets of culture and adapted them over time.*[20]

Well known from ancient Greek and Roman writers is King Midas, the last king of the Phrygian Kingdom. The mythology of Midas revolves around his ability to turn objects to gold by mere touch, as granted by Dionysos, and his unfortunate encounter with Apollo from which his ears are turned into the ears of a donkey. The historical record of Midas shows that he lived approximately between 740 and 696 BCE, and represented Phrygia as a great king. Most historians now consider him to be King Mita of the Mushkis as noted in Assyrian accounts. The Assyrians thought of Mita as a dangerous foe, for Sargon II, their ruler at the time, was quite happy to negotiate a peace treaty in 709 BCE. This treaty had no effect on the advancing Cimmerians in the East, who streamed into Phrygia and led to the downfall and suicide of King Midas in 696 BCE.*[21]

Following Midas's death Phrygia lost its independence, becoming respectively a vassal state of its western neighbour, Lydia, Persia, Greece, Rome and Byzantium, disappearing in the Turkish era.

7.4.3 Eastern Anatolia

Cimmeria

Main article: Cimmerians

 Cimmeria was a region of north eastern Anatolia, appearing in the 8th century BC from the north and east, in the face of the eastern Scythian advance. They continued to move west, invading and subjugating Phrygia (696-695 BC), penetrating as far south as Cilicia, and west into to Ionia after pillaging Lydia. Lydian campaigns between 637 and 626 BC effectively halted this advance. The Cimmerian influence progressively weakened and the last recorded mention is in 515 BC.

Urartu

Main article: Urartu

 Urartu (Nairi, or the Kingdom of Van) existed in north-east Anatolia, centered around Lake Van (Nairi Sea), to the south of the Cimmerians and North of Assyria. Its prominence ran from its appearance in the 9th century until it was overrun by the Medes in the 6th century.

Urartu is first mentioned as a loose confederation of smaller entities in the Armenian Highlands in the 13th to 11th centuries BC, but was subject to recurrent Assyrian incursions before emerging as a powerful neighbour by the 9th century BC. This was facilitated by Assyria's weak position in the 8th century BC. Urartu continued to resist Assyrian attacks and reached it greatest extent under Argishti I (c. 785–760 BC). At that time it included present day Armenia, southern Georgia reaching almost to the Black Sea, west to the sources of the Euphrates and south to the sources of the Tigris.

Following this Urartu suffered a number of setbacks. King Tiglath Pileser III of Assyria conquered it in 745 BC. By 714 BC it was being ravaged by both Cimmerian and Assyrian raids. After 645 BC Scythian attacks provided further problems for Urartu forcing it become dependent on Assyria. However Assyria itself fell to a combined attack of Scythians, Medes and Babylonians in 612 BC. While the details of Urartu's demise are debated, it effectively disappeared to be replaced by Armenia. It was a Persian Satrapy for a while from the 6th century BC before becoming an independent Armenia. To this day Urartu forms an important part of Armenian nationalist sentiment.

Assyria

Main article: Assyria

In the Iron Age Assyria extended to include south eastern Anatolia. Assyria, one of the great powers of the Mesopotamia region, had a long history from the 25th century BC (Bronze Age) until it final collapse in 612 BC at the end of the Iron Age. Assyria's Iron Age corresponds to the Middle Period (resurgence) and the Neo-Assyrian Empire in its last 300 years, and its territory centered on what is modern day Iraq.

Assyria influenced Anatolian politics and culture from when its traders first came into contact with Hattians in the late Bronze Age. By the 13th century BC Assyria was expanding to its north west at the expense of the Hittites, and to the north at the expense of Urartu. Assyrian expansion reached its height under Tukulti-Ninurta I (1244 BC - 1208 BC), following which it was weakened by internal dissent. The collapse of the Hittie Empire at the end of the Bronze Age coincided with an era of renewed Assyrian expansion under Ashur-resh-ishi I (1133 BC - 1116 BC) and soon Assyria had added the Anatolian lands in what is now Syria to its empire. Tiglath-Pileser I (1115 BC - 1077 BC) then commenced incursions against the Neo-Hittite Phrygians, followed by the Luwian kingdoms of Commagene, Cilicia and Cappadocia.

With the death of Tiglath-Pileser I Assyria entered a period of decline during what is referred to as the Ancient Dark Ages (1075-912 BC)in the region that corresponded to the collapse of the Bronze Age. The last 300 years of the Assyrian Empire (Neo-Assyrian Empire) from 911-627 BC saw a renewed expansion including attacks on the Neo-Hittite states to its north and west. Ashurnasirpal II (883–859 BC) extracted tribute from Phrygia while his successor Shalmaneser III (858–823 BC) also attacked Urartu forcing his Anatolian neighbours to pay tribute. After his death the land was torn by civil war. Assyrian power continued to wax and wane with periodic incursions into the Anatolian lands. Sennacherib (705-681 BC) encountered and drove back a new force in the region, the Greeks who attempted to settle Cilicia. His successor Esarhaddon (680-669 BC) was responsible for the final destruction of Urartu. Ashurbanipal (669-627 BC) then extended Assyrian influence still further placing Caria, Cilicia, Lydia and Cappadocia into vasselage.

However Assyria found its resources stretched to maintain the integrity of its vast empire and civil war again erupted following the death of Ashurbanipal. Vassal states stopped paying tribute, regaining independence. The weakened Assyrian state was now faced by a new threat, a coalition of Iranian peoples to its east and north, including Medes, Persians, Scythians and the Anatolian Cimmerians, who attacked Assyria in 616 BC. Ninevah, the capital, fell in 612 BC and the Assyrian Empire was finally swept away in 605 BC.

With the collapse of Assyria, ended not only the Iron Age, but also the era referred to as Pre-History, to make way for what has been variously described as Recorded History, or more specifically late Ancient History or Classical Civilisation. However these terms are not precise or universal and overlap.

7.5   BTC pipeline archaeological sites

Archaeological sites in Güllüdere, Yüceören, and Ziyaretsuyu were revealed by the BTC pipeline construction.*[22]

7.6   See also

- History of Europe
- History of the Middle East
- Timeline of Middle Eastern history
- Anatolianism

- Ancient kingdoms of Anatolia

- Ancient Regions of Anatolia

- Ancient Near East

7.7 Notes

[1] http://www.sci-news.com/archaeology/science-stone-tool-turkey-02370.html

[2] Suthan 2009-2014, Paleolithic age

[3] Manisa Museum, Republic of Turkey Culture minister website

[4] Martin Lockley, Gordon Roberts & Jeong Yul Kim. In the Footprints of Our Ancestors: An Overview of the Hominid Track Record. Ichnos Volume 15, Issue 3-4, 2008, pages 106-125

[5] Suthan 2009-2014, Mesolithic age

[6] Suthan 2009-2014, Neolithic age

[7] Suthan 2009-2014, Chalcolithic age

[8] A Brief History of Hattusha/Boğazköy

[9] The History Files: Hatti (Hattusa)

[10] Expedition in Ancient Anatolia: Hattians - First Civilizations in Anatolia

[11] The Joukowsky Institute of Archaeology: The Archaeology of Mesopotamia

[12] The Hittites, their forerunners and their followers

[13] Burney CA. Historical dictionary of the Hittites: Kültepe. Scarecrow Press, 2004, Lanham MD

[14] Saggs, H.W.F. (2000). *Babylonians.* University of California Press. ISBN 0-520-20222-8.

[15] Freeman, Charles (1999). *Egypt, Greece and Rome: Civilizations of the Ancient Mediterranean.* Oxford University Press. ISBN 0-19-872194-3.

[16] Trevor Bryce, *The Kingdom of the Hittites,* rev. ed, 2005:9.

[17] C. Michael Hogan, *Knossos fieldnotes*, Modern Antiquarian (2007)

[18] Hawkins, John David (2000). *Corpus of Hieroglyphic Luwian Inscriptions.* Walter de Gruyter. ISBN 3-11-014870-6.

[19] Duncker, Max (1879). *The History of Antiquity, Volume III.* Richard Bentley & Son.

[20] Garance Fiedler. "Phrygia" . Retrieved 2007-10-19.

[21] Encyclopædia Britannica Online. "The legends and the truth about King Midas" . Retrieved 2007-10-19.

[22] "Ancient Heritage in the BTC-SCP Pipeline Corridor" . *Smithsonian.* Retrieved 21 Apr 2014.

7.8 References

- Cambridge Ancient History Online 14 vols. 1970-2000
  Note: The original 11 vol Cambridge Ancient History 1928-36 is now available as free ebooks

- Cambridge Companions to the Ancient World. 10 vols.

- Duncker, Max (1879). *The History of Antiquity, Volume III*. Richard Bentley & Son.

- Hornblower, Simon; Antony Spawforth (1996). *The Oxford Classical Dictionary*. Oxford University Press.

- McEvedy, Colin (1967). *The Penguin Atlas of Ancient History*. Penguin.

- Marek, Christian (2010), *Geschichte Kleinasiens in der Antike* C. H. Beck, Munich, ISBN 978-3-406-59853-1 (review: M. Weiskopf, Bryn Mawr Classical Review 2010.08.13).

- Suthan, Resat (2009–2014). "Historical". *Anatolia*. Thracian Ltd.

Ancient Greek settlements in western Anatolia (11th-8th centuries BC). Halikarnassus was initially Dorian, then Ionian. Smyrna changed from Aeolian to Ionian

Lydian electrum coin, depicting a lion and bull.

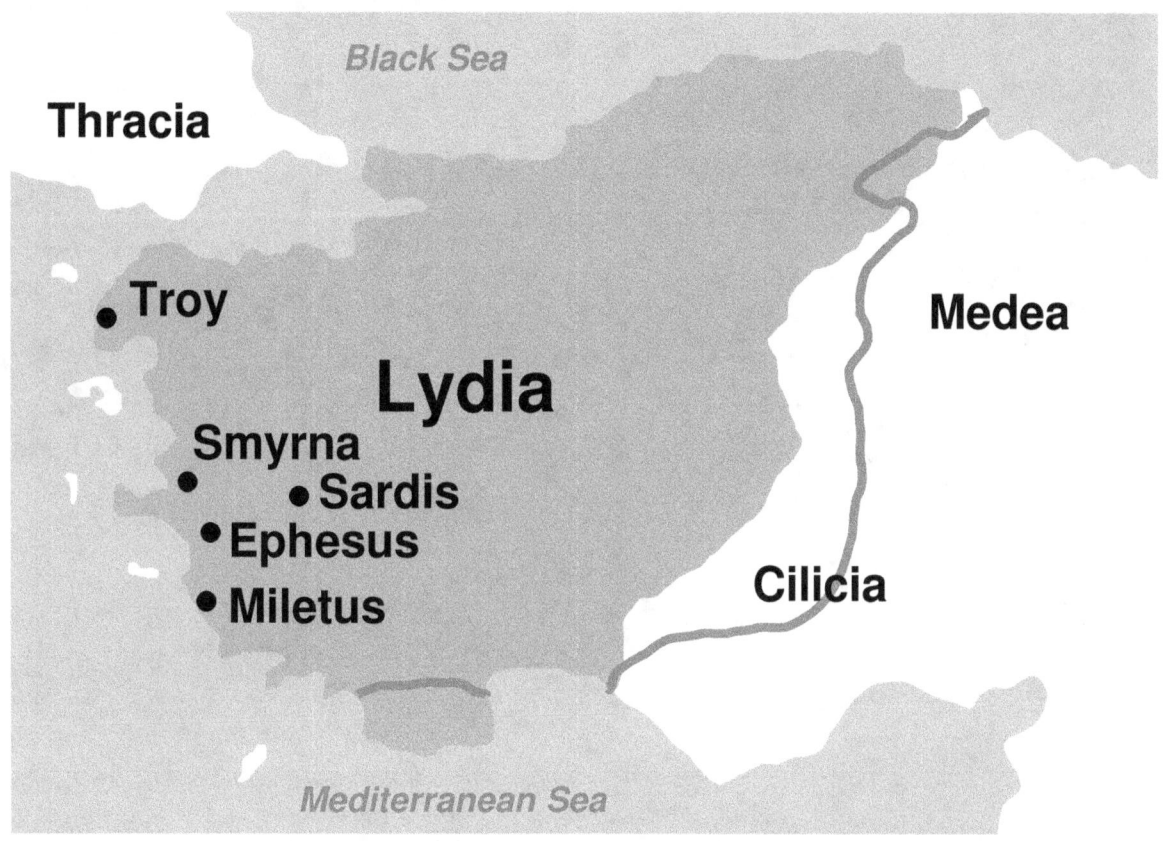

Map of the Lydian Empire under Croesus, 6th century BC

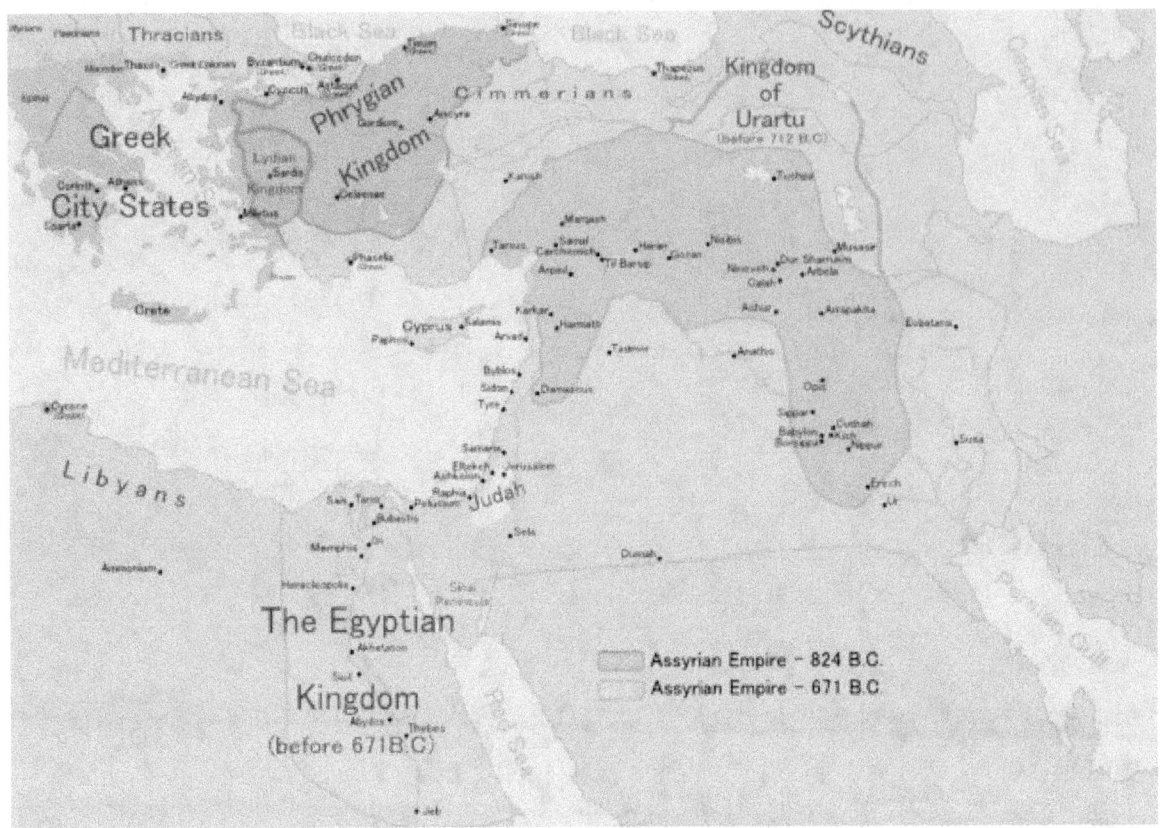

Lydia, Phrygia, Cimmeria and Assyria, 9th-7th centuries BCE

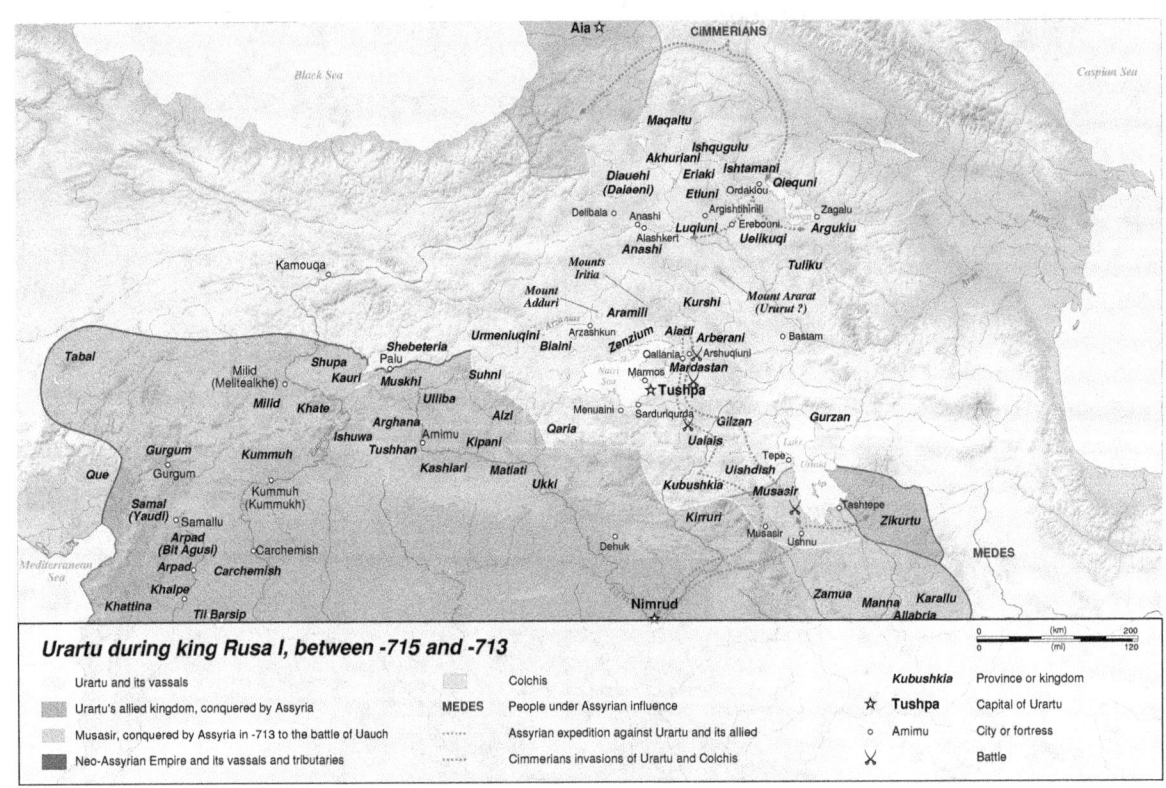

Cimmerian invasions of Colchis, Urartu and Assyria 715-3 BC

The Kingdom of Urartu, 9th-6th Centuries B.C.

Urartu 9th-6th centuries BC

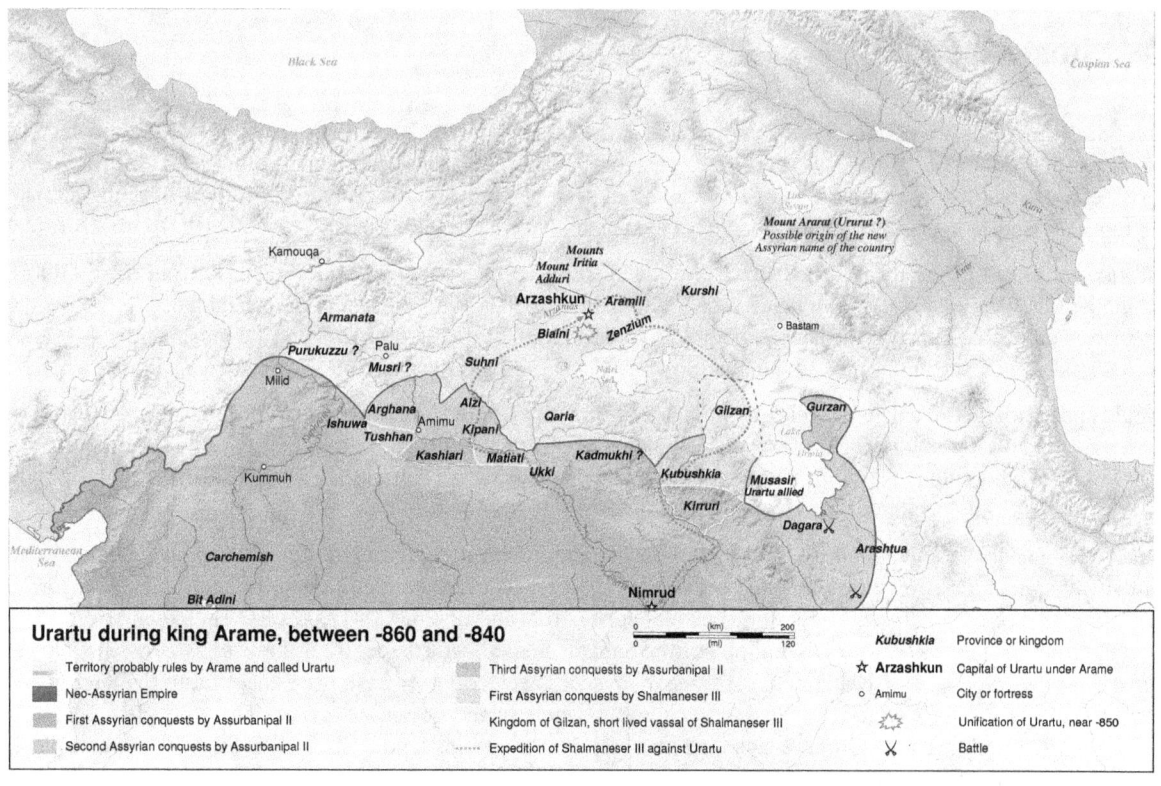

Urartu under Aramu 860-40 BC

Chapter 8

History of the ancient Levant

This article is about History of the Levant. For what the area is called by natives and others, see Names of the Levant. For region's history, see History of the Middle East.

The Levant is a geographical term that refers to a large area in Southwest Asia, south of the Taurus Mountains, bounded by the Mediterranean Sea in the west, the Arabian Desert in the south, and Mesopotamia in the east. It stretches 400 miles north to south from the Taurus Mountains to the Sinai desert, and 70 to 100 miles east to west between the sea and the Arabian desert.*[1] The term is also sometimes used to refer to modern events or states in the region immediately bordering the eastern Mediterranean Sea: Cyprus, Israel, Palestine, Jordan, Lebanon, and Syria.

The term normally does not include Anatolia (although at times Cilicia may be included), the Caucasus Mountains, Mesopotamia or any part of the Arabian Peninsula proper. The Sinai Peninsula is sometimes included, though it is more considered an intermediate, peripheral or marginal area forming a land bridge between the Levant and northern Egypt.

8.1 Stone Age

Anatomically modern Homo sapiens are demonstrated at the area of Mount Carmel, during the Middle Paleolithic dating from about c. 90,000 BC. This move out of Africa seems to have been unsuccessful and by c. 60,000 BC in Palestine/Israel/Syria, especially at Amud, classic Neanderthal groups seem to have profited from the worsening climate to have replaced Homo sapiens, who seem to have been confined once more to Africa.*[2]

A second move out of Africa is demonstrated by the Boker Tachtit Upper Paleolithic culture, from 52–50,000 BC, with humans at Ksar Akil XXV level being modern humans.*[3] This culture bears close resemblance to the Badoshan Aurignacian culture of Iran, and the later Sebilian I Egyptian culture of c. 50,000 BC. Stephen Oppenheimer*[4] suggests that this reflects a movement of modern human (possibly Caucasian) groups back into North Africa, at this time.

It would appear this sets the date by which Homo sapiens Upper Paleolithic cultures begin replacing Neanderthal Levalo-Mousterian, and by c. 40,000 BC Palestine was occupied by the Levanto-Aurignacian Ahmarian culture, lasting from 39–24,000 BC.*[5] This culture was quite successful spreading as the Antelian culture (late Aurignacian), as far as Southern Anatolia, with the Atlitan culture.

After the Late Glacial Maxima, a new Epipaleolithic culture appears in Southern Palestine. Extending from 18–10,500 BC, the Kebaran culture*[6] shows clear connections to the earlier Microlithic cultures using the bow and arrow, and using grinding stones to harvest wild grains, that developed from the c. 24,000–17,000 BC Halfan culture of Egypt, that came from the still earlier Aterian tradition of the Sahara. Some linguists see this as the earliest arrival of Nostratic languages in the Middle East. Kebaran culture was quite successful, and may have been ancestral to the later Natufian culture (10,500–8500 BC), which extended throughout the whole of the Levantine region. These people pioneered the first sedentary settlements, and may have supported themselves from fishing, and from the harvest of wild grains plentiful in the region at that time.

Natufian culture also demonstrates the earliest domestication of the dog, and the assistance of this animal in hunting and

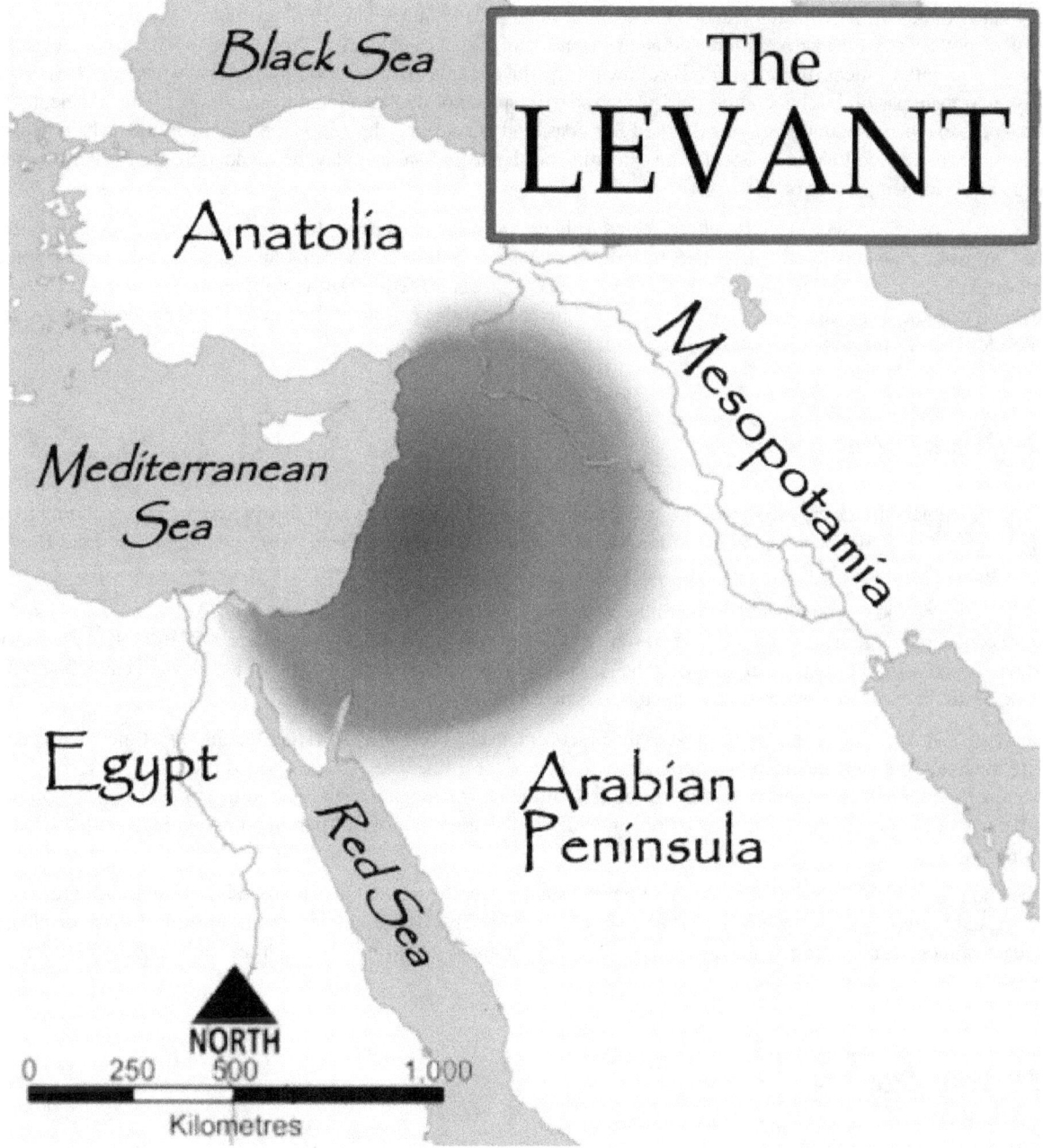

The Levant

guarding human settlements may have contributed to the successful spread of this culture. In the northern Syrian, eastern Anatolian region of the Levant, Natufian culture at Cayonu and Mureybet developed the first fully agricultural culture with the addition of wild grains, later being supplemented with domesticated sheep and goats, which were probably domesticated first by the Zarzian culture of Northern Iraq and Iran (which like the Natufian culture may have also developed from Kebaran).

By 8500–7500 BC, the Pre-Pottery Neolithic A (PPNA) culture developed out of the earlier local tradition of Natufian in Southern Palestine, dwelling in round houses, and building the first defensive site at Jericho (guarding a valuable fresh water spring). This was replaced in 7500 BC by Pre-Pottery Neolithic B (PPNB), dwelling in square houses, coming from Northern Syria and the Euphrates bend.

During the period of 8500–7500 BC, another hunter-gatherer group, showing clear affinities with the cultures of Egypt (particularly the Outacha retouch technique for working stone) was in Sinai. This Harifian culture[7] may have adopted the use of pottery from the Isnan culture and Helwan culture of Egypt (which lasted from 9000 to 4500 BC), and subsequently fused with elements from the PPNB culture during the climatic crisis of 6000 BC to form what Juris Zarins calls the Syro-Arabian pastoral technocomplex,[8] which saw the spread of the first Nomadic pastoralists in the Ancient Near East. These extended southwards along the Red Sea coast and penetrating the Arabian bifacial cultures, which became progressively more Neolithic and pastoral, and extending north and eastwards, to lay the foundations for the tent-dwelling Martu and Akkadian peoples of Mesopotamia.

In the Amuq valley of Syria, PPNB culture seems to have survived, influencing further cultural developments further south. Nomadic elements fused with PPNB to form the Minhata Culture and Yarmukian Culture which were to spread southwards, beginning the development of the classic mixed farming Mediterranean culture, and from 5600 BC were associated with the Ghassulian culture of the region, the first chalcolithic culture of the Levant. This period also witnessed the development of megalithic structures, which continued into the Bronze Age.[9]

8.2   Bronze Age

In modern scholarship the chronology of the Bronze age Levant is divided into Early/Proto Syrian, corresponding to the Early Bronze; Old Syrian, corresponding to the Middle Bronze; and Middle Syrian, corresponding to the Late Bronze. The term Neo-Syria is used to designate the early Iron Age.[10]

The old Syrian period was dominated by the Eblaite first kingdom, Nagar and the Mariote second kingdom. The Akkadian Empire conquered large areas of the Levant and were followed by the Amorite kingdoms, ca. 2000–1600 BC, which arose in Mari, Yamkhad and Qatna. Also following the Akkadians was the extension of Khirbet Kerak ware culture, showing affinities with the Caucasus, and possibly linked to the later appearance of the Hurrians.

Around the 17th and 16th centuries BC most of the older centers had been overrun. The Mitanni, for a time, menaced the Hittite kingdom, but were defeated by it around the middle of the 14th. The Semitic Hyksos used the new technologies to occupy Egypt, but were expelled, leaving the empire of the New Kingdom to develop in their wake. From 1550 until 1100, much of the Levant was conquered by Egypt, which in the latter half of this period contested Syria with the Hittite Empire.

At the end of the 13th century BC, all of these powers suddenly collapsed. Cities all around the eastern Mediterranean were sacked within a span of a few decades by assorted raiders.The Hittite empire was destroyed. Egypt repelled its attackers with only a major effort, and over the next century shrank to its territorial core, its central authority permanently weakened.

8.3   Iron Age

The destruction at the end of the Bronze Age left a number of tiny kingdoms and City-states behind. A few Hittite centres remained in northern Syria, along with some Phoenician ports in Canaan that escaped destruction and developed into great commercial powers. The Israelites emerged as a rural culture (possibly from the displaced Canaanite refugees escaping the Bronze Age Collapse to Judea and Samaria alongside groups like the Shasu and the Habiru) mainly in the Canaanite hill-country and the Eastern Galilee, quickly spreading through the land and forming an alliance in the struggle for the land against the Philistines to the West, Moab and Ammon to the East and Edom to the South. In the 12th century BC, most of the interior, as well as Babylonia, was overrun by Arameans, while the shoreline around today's Gaza Strip was settled by Philistines.

In this period a number of technological innovations spread, most notably iron working and the Phoenician alphabet, developed by the Phoenicians or the Canaanites around the 16th century BC.

During the 9th century BC, the Assyrians began to reassert themselves against the incursions of the Aramaeans, and over the next few centuries developed into a powerful and well-organised empire. Their armies were among the first to employ cavalry, which took the place of chariots, and had a reputation for both prowess and brutality. At their height, the Assyrians dominated all of the Levant, Egypt, and Babylonia. However, the empire began to collapse toward the end of the 7th century BC, and was obliterated by an alliance between a resurgent New Kingdom of Babylonia and the Iranian Medes.

The subsequent balance of power was short-lived, though. In the 550s BC the Persians revolted against the Medes and gained control of their empire, and over the next few decades annexed to it the realms of Lydia in Anatolia, Damascus, Babylonia, and Egypt, as well as consolidating their control over the Iranian plateau nearly as far as India. This vast kingdom was divided up into various satrapies and governed roughly according to the Assyrian model, but with a far lighter hand. Around this time Zoroastrianism became the predominant religion in Persia.

8.4 Classical Age

Persia controlled the Levant but by the 4th century BC, Persia had fallen into decline. The campaigns of Xenophon illustrated how very vulnerable Persia had become to attack by an army organized along Greek lines, and under Alexander the Great the Levant was conquered.

Alexander did not live long enough to consolidate his realm, the greater share of the east went to the descendants of Seleucus I Nicator. This period saw great innovations in mathematics, science, architecture, and the like, and Greeks founded cities throughout the east, some of which grew to be the world's first major metropolises. Their culture did not, however, reach very far into the countryside.

The Seleucids adopted a pro-western stance that alienated both the powerful eastern satraps and the Greeks who had migrated to the east. During the 2nd century BC, Greek culture lost ground there, and the empire began to break apart. The Seleucid kingdom continued to decline and its remaining provinces were annexed by the Roman Republic in 64 BC as Iudaea Province.

Persian dynasty, the Sassanids, entered in conflicts with Rome, and later with the Byzantine Empire. In 391, the Byzantine era began with the permanent division of the Roman Empire into East and Western halves. Byzantine control over the sites of Israel and Judah and other parts of the Levant lasted until 636, when it was conquered by Arabs and became a part of the Caliphate.

The Byzantines reached their lowest point under Phocas, with the Sassanids occupying the whole of the eastern Mediterranean. In 610, though, Heraclius took the throne of Constantinople and began a successful counter-attack, expelling the Persians and invading Media and Assyria. Unable to stop his advance, Khosrau II was assassinated and the Sassanid empire fell into anarchy. Weakened by their quarrels, neither empire was prepared to deal with the onslaught of the Arabs, newly unified under the banners of Islam and anxious to expand their faith. By 650, Arab forces had conquered all of Persia, Syria, and Egypt.

8.5 Later eras

Further information: History of Islam
Further information: History of the Middle East

8.6 See also

- Ancient Near East

- History of the Northern Levant

  - History of Cyprus

- History of the Southern Levant

  - History of Palestine (region)

- History of the Southeastern Levant
- Pre-history of the Southern Levant
- History of ancient Israel and Judah
- History of the Sinai Peninsula

8.7 References

8.7.1 Notes

[1] *A History of Ancient Israel and Judah* by Miller, James Maxwell, and Hayes, John Haralson (Westminster John Knox, 1986) ISBN 0-664-21262-X. p.36]

[2] "Amud". *Encyclopædia Britannica*. Retrieved 2007-10-11.

[3] Marks, Anthony (1983)"Prehistory and Paleoenvironments in the Central Negev, Israel" (Institute for the Study of Earth and Man, Dallas)

[4] Oppemheiomer, Stephen (2004), "Out of Eden", (Constable and Robinson)

[5] Gladfelter, Bruce G. (1997) "The Ahmarian tradition of the Levantine Upper Paleolithic: the environment of the archaeology" (Vol 12, 4 *Geoarchaeology*)

[6] Ronen, Avram, "Climate, sea level, and culture in the Eastern Mediterranean 20 ky to the present" in Valentina Yanko-Hombach, Allan S. Gilbert, Nicolae Panin and Pavel M. Dolukhanov (2007), *The Black Sea Flood Question: Changes in Coastline, Climate, and Human Settlement* (Springer)

[7] Belfer-Cohen, Anna and Bar-Yosef, Ofer "Early Sedentism in the Near East: A Bumpy Ride to Village Life" (*Fundamental Issues in Archaeology*, 2002, Part II, 19–38)

[8] Zarins, Yuris "Early Pastoral Nomadiism and the Settlement of Lower Mesopotamia" (# *Bulletin of the American Schools of Oriental Research* No. 280, November, 1990)

[9] Scheltema, H.G. (2008). *Megalithic Jordan: An Introduction and Field Guide*. Amman, Jordan: The American Center of Oriental Research. ISBN 978-9957-8543-3-1

[10] Mogens Herman Hansen (2000). *A Comparative Study of Thirty City-state Cultures: An Investigation, Volume 21*. p. 57.

8.7.2 General references

- Philip Mansel, *Levant: Splendour and Catastrophe on the Mediterranean*, London, John Murray, 11 November 2010, hardback, 480 pages, ISBN 978-0-7195-6707-0, New Haven, Yale University Press, 24 May 2011, hardback, 470 pages, ISBN 978-0-300-17264-5

8.8 External links

- The History of the Ancient Near East

Chapter 9

Gojoseon

Gojoseon (Korean pronunciation: [kodʑosʰʌn]) was an ancient Korean kingdom. The addition of *Go* (고, 古), meaning "ancient", distinguishes it from one of the various names of Joseon.

The founding legend of Gojoseon, which is recorded in the *Samguk Yusa* (1281) and other medieval Korean books,[*][1] states that the country was established in 2333 BC by Dangun, said to be descended from heaven.[*][2] While no evidence has been found that supports whatever facts may lie beneath this,[*][3] the account has played an important role in developing Korean national identity.

In the 12th century BC Gija, a prince from the Shang dynasty of China, purportedly founded Gija Joseon. However, due to the lack of archaeological evidence, its existence has been challenged since the 20th century, and today no longer forms the mainstream understanding of this period.

The historical Gojoseon kingdom was first mentioned in Chinese records in the early 7th century BC.[*][4] During its early phase, the capital of Gojoseon was located in Liaoning; around 400 BC, and was moved to Pyongyang, while in the south of the peninsula, the Jin state arose by the 3rd century BC.[*][5]

In 108 BC, the Han dynasty of China invaded and conquered Wiman Joseon. The Han established four commanderies to administer the Gojoseon territory. The area was later conquered by Goguryeo in 313 AD.

Their language was probably a predecessor of the equally prehistoric Buyeo languages, and perhaps a form of Proto-Korean.[*][6]

9.1 Founding myth

Main article: Korean mythology § Founding myth

9.1.1 Dangun Myth and its controversy

Dangun is the legendary founder of Korea. While there are variations among different texts and oral traditions, the most popular account is from the written record of the founding myth in the *Samgungnyusa,* a 13th-century collection of legends and stories. A similar account is found in *Jewang Ungi*. According to the legend, the Lord of Heaven, Hwanin had a son, Hwanung, who descended to Baekdu Mountain and founded the city of Shinsi. Hwanung later married a bear turned into a woman, Ungnyeo (웅녀, 熊女), and Ungnyeo gave birth to Dangun.

While the Dangun story is considered to be a myth,[*][7] it is believed it is a mythical synthesis of a series of historical events relating to the founding of Gojoseon.[*][8] There are various theories on the origin of this myth.[*][9] Seo and Kang (2002) believe the Dangun myth is based on integration of two different tribes, an invasive sky-worshipping Bronze Age tribe and a native bear-worshipping neolithic tribe, that led to the foundation of Gojoseon.[*][10] Lee K. B. (1984) believes

Heaven Lake of Baekdu Mountain, where Dangun's father is said to have descended from heaven

'Dangun-wanggeom' was a title borne by successive leaders of Gojoseon.[*][11]

Dangun is said to have founded Gojoseon around 2333 BC, based on the descriptions of the *Samgungnyusa*, *Jewang Ungi*, *Dongguk Tonggam* and the *Annals of the Joseon Dynasty*.[*][12] The date differs among historical sources, although all of them put it during the mythical Emperor Yao's reign (traditional dates: 2357 BCE? – 2256 BCE?). *Samgungnyusa* says Dangun ascended to the throne in the 50th year of the legendary Yao's reign, *Annals of the King Sejong* says the first year, and *Dongguk Tonggam* says the 25th year.[*][13]

9.1.2   Gija Myth and its controversy

Main article: Gija Joseon

Gija, a man from Shang China, allegedly fled to the Korean peninsula in 1122 BC during the fall of the Shang to the Zhou dynasty and founded Gija Joseon.[*][14] In South Korea, the United Kingdom and the United States, most scholars believe Gija's relation to Gojoseon is a Chinese fabrication and Gija has nothing to do with Gojoseon.[*][9] In the past, the earliest surviving Korean record, *Records of the Three Kingdoms*, admitted Gija Joseon. The *Dongsa Gangmok* of 1778 described Gija's activities and contributions in Gojoseon. The records of Gija refer to eight laws (Hangul: 범금팔조; hanja: 犯禁八條), that are recorded by the *Book of Han* and evidence a hierarchical society and legal protection of private property.[*][15]

In pre-modern Korea, Gija represented the authenticating presence of Chinese civilization, and until the 12th century Koreans commonly believed that Danjun bestowed upon Korea its people and basic culture, while Gija gave Korea its high culture—and presumably, standing as a legitimate civilization.[*][16]

However, in the modern era Gija's place has diminished to the point of near extinction.[16] Many Korean scholars deny its existence for various reasons, mainly due to contradicting archaeological evidence.[17] They also point to the *Bamboo Annals* and the *Analects* of Confucius, which were among the first works to mention Gija, but do not mention his migration to Gojoseon.[18]

9.2 State formation

See also: Gojoseon–Yan War

Gojoseon is first found in contemporaneous historical records of early 7th century BC, as located around Bohai Bay and trading with Qi (齊) of China.[19] The *Zhan Guo Ce*, *Classic of Mountains and Seas*, and *Records of the Grand Historian* refers to Joseon as a region until the *Records of the Grand Historian* began referring it as a country from 195 BCE onwards.[14]

By the 4th century BCE, other states with defined political structures developed in the areas of the earlier Bronze Age "walled-town states"; Gojoseon was the most advanced of them in the peninsular region.[5] The city-state expanded by incorporating other neighboring city-states by alliance or military conquest. Thus, a vast confederation of political entities between the Taedong and Liao rivers was formed. As Gojoseon evolved, so did the title and function of the leader, who came to be designated as "king" (Han), in the tradition of the Zhou dynasty, around the same time as the Yan (燕) leader.[20] Records of that time mention the hostility between the feudal state in Northern China and the "confederated" kingdom of Gojoseon, and notably, a plan to attack the Yan beyond the Liao River frontier. The confrontation led to the decline and eventual downfall of Gojoseon, described in Yan records as "arrogant" and "cruel". But the ancient kingdom also appears as a prosperous Bronze Age civilization, with a complex social structure, including a class of horse-riding warriors who contributed to the development of Gojoseon, particularly the northern expansion[21] into most of the Liaodong basin.

Around 300 BC, Gojoseon lost significant western territory after a war with the Yan state, but this indicates Gojoseon was already a large enough state that could wage war against Yan and survive the loss of 2000 li (800 kilometers) of territory.[15] Gojoseon is thought to have relocated its capital to the Pyongyang region around this time.[20]

9.2.1 Wiman Joseon and fall

See also: Wiman Joseon, Gojoseon–Han War and Four Commanderies of Han

In 195 BCE, Jun of Gojoseon appointed a refugee from Yan, Wiman.[22] Wiman later rebelled in 194 BCE, and Jun fled to the south of the Korean Peninsula.[23]

In 109 BCE, Emperor Wu of Han invaded near the Liao River.[23] A conflict would erupt in 109 BCE, when Wiman's grandson King Ugeo (□□□, hanja: 右渠王) refused to permit Jin's ambassadors to reach China through his territories. When Emperor Wei sent an ambassador She He (涉何) to Wanggeom-seong to negotiate right of passage with King Ugeo, King Ugeo refused and had a general escort She back to Han territory —but when they got close to Han borders, She assassinated the general and claimed to Emperor Wu that he had defeated Joseon in battle, and Emperor Wu, unaware of his deception, made him the military commander of the Commandery of Liaodong. King Ugeo, offended, made a raid on Liaodong and killed She He.

In response, Emperor Wu commissioned a two-pronged attack, one by land and one by sea, against Joseon.[23] The two forces attacking Joseon were unable to coordinate well with each other and eventually suffered large losses. Eventually the commands were merged, and Wanggeom fell in 108 BC. Han took over the Joseon lands and established Four Commanderies of Han in the western part of former Gojoseon area.[24]

The Gojoseon disintegrated by 1st century BC as it gradually lost the control of its former fiefs. As Gojoseon lost control of its confederacy, many successor states sprang from its former territory, such as Buyeo, Okjeo, Dongye. Goguryeo and Baekje evolved from Buyeo.

9.3 Culture

Around 2000 BCE, a new pottery culture of painted and chiseled design is found. These people practiced agriculture in a settled communal life, probably organized into familial clans. Rectangular huts and increasingly larger dolmen burial sites are found throughout the peninsula. Bronze daggers and mirrors have been excavated, and there is archaeological evidence of small walled-town states in this period.[21][25] Dolmens and bronze daggers found in the area are uniquely Korean and can't be found in China.

9.3.1 Mumun pottery

In the Mumun pottery period (1500–300 BCE), plain coarse pottery replaced earlier comb-pattern wares, possibly as a result of the influence of new populations migrating to Korea from Manchuria and Siberia. This type of pottery typically has thicker walls and displays a wider variety of shapes, indicating improvements in kiln technology.[5] This period is sometimes called the "Korean bronze age", but bronze artifacts are relatively rare and regionalized until the 7th century BCE.

9.3.2 Rice cultivation

Sometime around 1200 to 900 BCE, rice cultivation spread to Korea from China and Manchuria. The people also farmed native grains such as millet and barley, and domesticated livestock.[26]

9.3.3 Bronze tools

Main article: Liaoning bronze dagger culture

The beginning of the Bronze Age on the peninsula is usually said to be 1000 BCE, but estimates range from the 13th to 8th centuries.[27] Although the Korean Bronze Age culture derives from the Liaoning and Manchuria, it exhibits unique typology and styles, especially in ritual objects.

By the 7th century BCE, a Bronze Age material culture with influences from Manchuria, eastern Mongolia as well as Siberia and Scythian bronze styles, flourished on the peninsula. Korean bronzes contain a higher percentage of zinc than those of the neighboring bronze cultures. Bronze artifacts, found most frequently in burial sites, consist mainly of swords, spears, daggers, small bells, and mirrors decorated with geometric patterns.[5][28]

Gojoseon's development seems linked to the adoption of bronze technology. Its singularity finds its most notable expression in the idiosyncratic type of bronze swords, or "mandolin-shaped daggers" (□□□□□, 琵琶形銅劍). The mandolin-shape dagger is found in the regions of Liaoning, Hebei, and Manchuria down to the Korean Peninsula. It suggests the existence of Gojoseon dominions. Remarkably, the shape of the "mandolin" dagger of Gojoseon differs significantly from the sword artifacts found in China.

9.3.4 Dolmen tombs

Megalithic dolmens appears in Korean peninsula and Manchuria around 2000 to 400 BCE.[29][30] Around 900 BC, burial practices become more elaborate, a reflection of increasing social stratification. Goindol, the dolmen tombs in Korea and Manchuria, formed of upright stones supporting a horizontal slab, are more numerous in Korea than in other parts of East Asia. Other new forms of burial are stone cists (underground burial chambers lined with stone) and earthenware jar coffins. The bronze objects, pottery, and jade ornaments recovered from dolmens and stone cists indicate that such tombs were reserved for the elite class.[5][31]

Around the 6th century BCE, burnished red wares, made of a fine iron-rich clay and characterized by a smooth, lustrous surface, appear in dolmen tombs, as well as in domestic bowls and cups.[5]

9.3.5 Iron culture

Main article: Jin (Korean state)

Around this time, the state of Jin occupied the southern part of the Korean peninsula. Very little is known about this state except it was the apparent predecessor to the Samhan confederacies.

Around 300 BCE, iron technology was introduced into Korea from Yan state. Iron was produced locally in the southern part of the peninsula by the 2nd century BCE. According to Chinese accounts, iron from the lower Nakdong River in the southeast was valued throughout the peninsula and Japan.*[5]

9.4 Proto–Three Kingdoms

Main article: Proto–Three Kingdoms of Korea

Numerous small states and confederations arose from the remnants of Gojoseon, including Goguryeo, the Buyeo kingdom, Jeon-Joseon, Okjeo, and Dongye. Three of the Chinese commanderies fell to local resistance within a few decades, but the last, Nakrang, remained an important commercial and cultural outpost until it was destroyed by the expanding Goguryeo in 313.

Jun of Gojoseon is said to have fled to the state of Jin in the southern Korean Peninsula. Jin developed into the Samhan confederacies, the beginnings of Baekje and Silla, continuing to absorb migration from the north. The Samhan confederacies were Mahan, Jinhan, and Byeonhan. King Jun ruled Mahan, which was eventually annexed by Baekje. Goguryeo, Baekje, and Silla gradually grew into the Three Kingdoms of Korea that dominated the entire peninsula by around the 4th century.

9.5 See also

- Names of Korea
- History of Korea

9.6 Notes

[1] See also *Jewang Ungi* (1287) and *Dongguk Tonggam* (1485).

[2] Hwang 2010, p. 2.

[3] Connor 2002, p. 10.

[4] Peterson & Margulies 2009, p. 6.

[5] "Timeline of Art and History, Korea, 1000 BC – 1 AD". Metropolitan Museum of Art.

[6] Jaehoon Lee (2004). "The Relatedness Between The Origin of Japanese and Korean Ethnicity" (PDF). The Florida State University. p. 31. Retrieved 2007-04-11.

[7] Seth, Michael J. (2010) History of Korea: From Antiquity to the Present, Rowman & Littlefield Publishers. ISBN 978-0-7425-6717-7.

[8] 고조선 (古朝鮮). *Encylopaedia Britannica(Korean)* (in Korean).

[9] Barnes 2001, pp. 9–14.

[10] 2002.

[11] Lee 1984.

[12] □□□□ 24 □□□□□ - □□□□□□□□□□□ - □□□□□□□□□□□ 《□□□□》 □□□□□□□ 25 □□□□□□□□□□□. 《□□□□》〈□□〉 □□□□□□□□□□□□□□.
      - 古記云, 檀君與堯竝立於戊辰, 虞夏至商武丁八年乙未, 入阿斯達山□神, 享壽千四百十八年. 此説可疑今按, 堯之立
      在上元甲子甲辰之歲, 而檀君之立在後二十五年戊辰, 則曰與堯竝立者非也. □□□□□□□□ □□□□□□□□□□□ (□□: □□□□□□
      □□) □□□□□□□□.

[13] Yoon, N.-H. (□□□), The Location and Transfer of Go-Chosun's Capital (□□□□□□□□□□□□), □□□□□, **7**, 207–38 (2002)

[14] Barnes 2001, pp. 9–10.

[15] (Korean) Daum □□□□: □□□

[16] Kyung Moon hwang, "A History of Korea, An Episodic Narrative", 2010, pp. 4

[17] http://www.dbpia.co.kr/view/ar_view.asp?pid=694&isid=30674&arid=657709&topMenu=&topMenu1=

[18] □□□□□□□

[19] 고조선 (in Korean). Naver/Doosan Encyclopedia.

[20] (Korean) http://100.naver.com/100.php?id=14543

[21] "Korea's Place in the Sun". *The New York Times*.

[22] Academy of Korean Studies, 〈The Review of Korean Studies〉, vol. 10 □,3-4, 2007, p.222

[23] Lee Injae, Owen Miller, Park Jinhoon, Yi Hyun-Hae, 《Korean History in Maps》, Cambridge University Press, 2014, p.20

[24] Jae-eun Kang, 《The Land of Scholars: Two Thousand Years of Korean Confucianism》, Homa & Sekey Books, 2006, p.28-31

[25] North Korea - The Origins Of The Korean Nation

[26] "Timeline of Art and History". Metropolitan Museum of Art.

[27] (Korean) □□□□□青銅器文化 (□□□□□, 2001.12, □□□□□□□□)

[28] The Metropolitan Museum of Art: Arts of Korea, Bronze Age Objects

[29] http://www.wontackhong.com/homepage1/data/1061.pdf

[30] https://books.google.com/books?id=rHeb7wQu0xIC&pg=PA79&dq=dolmen+in+korea+2000+BCE&hl=en&sa=X&ei=zfVA
      ved=0CB0Q6AEwAA#v=onepage&q=dolmen%20in%20korea%202000%20BCE&f=false

[31] Unesco.

9.7 Bibliography

- Barnes, Gina Lee (2001). *State Formation in Korea: Historical and Archaeological Perspectives*. Psychology Press. ISBN 978-0-7007-1323-3.

- Lee, Ki-Baik (1984). *A New History of Korea*. Harvard University Press. ISBN 0-674-61575-1.

- □, □□; □, □□ (2002). 뿌리깊은한국사샘이깊은이야기 1 : 고조선 · 삼국 [*Deep-rooted Korean History 1 : Gojoseon · Three Kingdoms*] (in Korean). □. ISBN 8981335362.

Chapter 10

Mumun pottery period

The **Mumun pottery period** is an archaeological era in Korean prehistory that dates to approximately 1500-300 BC[*]
This period is named after the Korean name for undecorated or plain cooking and storage vessels that form a large part of the pottery assemblage over the entire length of the period, but especially 850-550 BC.

The Mumun period is known for the origins of intensive agriculture and complex societies in both the Korean Peninsula and the Japanese Archipelago.[*][2][*][3][*][4] This period or parts of it have sometimes been labelled as the "Korean Bronze Age", after Thomsen's 19th century Three-age system classification of human prehistory. However, the application of such terminology in the Korean case is misleading since local bronze production did not occur until approximately the late 8th century BC at the earliest, bronze artifacts are rare, and the distribution of bronze is highly regionalized until after 300 BC.[*][5][*][6] A boom in the archaeological excavations of Mumun Period sites since the mid-1990s has recently increased our knowledge about this important formative period in the prehistory of East Asia.

The Mumun period is preceded by the Jeulmun Pottery Period (c. 8000-1500 BC). The Jeulmun was a period of hunting, gathering, and small-scale cultivation of plants.[*][6] The origins of the Mumun Period are not well known, but the megalithic burials, Mumun pottery, and large settlements found in the Liao River Basin and North Korea c. 1800-1500 probably indicate the origins of the Mumun Period of Southern Korea. Slash-and-burn cultivators who used Mumun pottery displaced people using Jeulmun Period subsistence patterns.[*][7]

10.1 Chronology

10.1.1 Early Mumun

The Early (or Formative) Mumun (c. 1500-850 BC) is characterized by shifting cultivation, fishing, hunting, and discrete settlements with rectangular semi-subterranean pit-houses. The social scale of Early Mumun societies was egalitarian in nature, but the latter part of this period is characterized by increasing intra-settlement competition and perhaps the presence of part-time "big-man" leadership.[*][8] Early Mumun settlements are relatively concentrated in the river valleys formed by tributaries of the Geum River in West-central Korea. However, one of the largest Early Mumun settlements, Eoeun (Hangeul: 어은), is located in the Middle Nam River valley in South-central Korea. In the latter Early Mumun, large settlements composed of many long-houses such as Baekseok-dong (Hangeul: 백석동) appeared in the area of modern Cheonan City, Chungcheong Nam-do.

Important long-term traditions related to Mumun ceremonial and mortuary systems originated in this sub-period. These traditions include the construction of megalithic burials, the production of red-burnished pottery, and production of polished groundstone daggers.

Ganghwa dolmen, South Korea

10.1.2   Middle Mumun

The Middle (or Classic) Mumun (c. 850-550 BC) is characterized by intensive agriculture, as evidenced by the large and expansive dry-field remains (c. 32,500 square metres) recovered at Daepyeong, a sprawling settlement with several multiple ditch enclosures, hundreds of pit-houses, specialized production, and evidence of the presence of incipient elites and social competition.*[2]*[3]*[9] A number of wet-field features have been excavated in southern Korea, indicating that paddy field rice-farming was also practiced.

Burials dating to the latter part of the Middle Mumun (c. 700-550 BC) contain a few high status mortuary offerings such as bronze artifacts. Bronze production probably began around this time in Southern Korea. Other high status burials contain greenstone (or jade) ornaments.*[4]*[9] A number of megalithic burials with deep shaft interments, substantial 'pavements' of rounded cobblestone, and prestige artifacts such as bronze daggers, jade, and red-burnished vessels were built in the vicinity of the southern coast in the Late Middle Mumun. High status megalithic burials and large raised-floor buildings at the Deokcheon-ni (Hangeul: □□□) and Igeum-dong sites in Gyeongsang Nam-do provide further evidence of the growth of social inequality and the existence of polities that were organized in ways that appear to be similar to simple "chiefdoms".*[4]

Korean archaeologists sometimes refer to Middle Mumun culture as *Songguk-ri* Culture (Hanja: 松菊里文化; Hangeul: □□□□□).*[1] Co-occurring artifacts and features that are grouped together as Songguk-ri Culture are found in settlement sites in the Hoseo and Honam regions of southeast Korea, but Songguk-ri Culture settlements are also found in western Yeongnam. Excavations have also revealed Songguk-ri settlements in the Ulsan and Gimhae areas. In 2005 archaeologists uncovered Songguk-ri Culture pit-houses at a site deep in the interior of Gangwon Province. The ultimate geographic reach of Songguk-ri Culture appears to have been Jeju Island and western Japan.

Mumun culture is the beginning of a long-term tradition of rice-farming in Korea that links Mumun Culture with the present-day, but evidence from the Early and Middle Mumun suggests that, although rice was grown, it was not the dominant crop.*[3] During the Mumun people grew millets, barley, wheat, legumes, and continued to hunt and fish.

Large Middle Mumun (c. 8th century BC storage vessel unearthed from a pit-house in or near Daepyeong, H= c. 60-70 cm.

10.1.3 Late Mumun

The Late (or Post-classic) Mumun (550-300 BC) is characterized by increasing conflict, fortified hilltop settlements, and a concentration of population in the southern coastal area. A Late Mumun occupation was found at the Namsan settlement, located on the top of a hill 100 m above sea level in modern Changwon City, Gyeongsang Nam-do. A shellmidden (shellmound) was found in the vicinity of Namsan, indicating that, in addition to agriculture, shellfish exploitation was part of the Late Mumun subsistence system in some areas. Pit-houses at Namsan were located inside a ring-ditch that is some 4.2 m deep and 10 m in width. Why would such a formidable ring-ditch, so massive in size, have been necessary? One possible answer is intergroup conflict. Archaeologists propose that the Late Mumun was a period of conflict between groups of people.

The number of settlements in the Late Mumun is much lower than in the previous sub-period. This indicates that populations were reorganized and settlement was probably more concentrated in a smaller number of larger settlements. There are a number of reasons why this could have occurred. There are some indications that conflict increased or climatic change led to crop failures.

Notably, according to the traditional Yayoi chronological sequence, Mumun-esque settlements appeared in Northern Kyūshū (Japan) during the Late Mumun. The Mumun period ends when iron appeared in the archaeological record along with pit-houses that had interior composite hearth-ovens reminiscent of the historic period (Hangeul: □□□, *agungi*).

Some scholars suggest that the Mumun pottery period should be extended to c. 0 BC because of the presence of an undecorated ware that was popular between 400 BC and 0 BC called *jeomtodae* (ko: □□□). However, bronze became very important in ceremonial and elite life from 300 BC. Additionally, iron tools are increasingly found in Southern Korea after 300 BC These factors clearly differentiate the time period 300 BC - 0 from the cultural, technological, and social

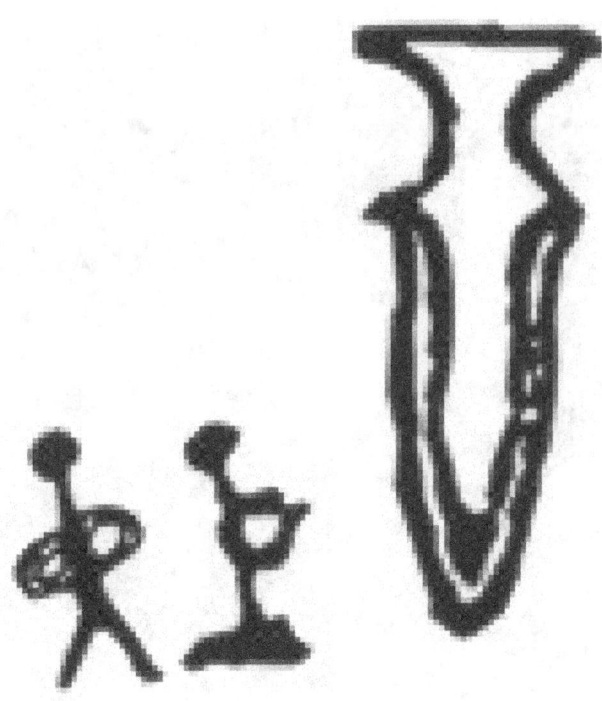

Representations of a dagger (right)and two human figures, one of which is kneeling (left), carved into the capstone of Megalithic Burial No. 5, Orim-dong, Yeosu, Korea.

scale that was present in the Mumun pottery period. The unequal presence of bronze and iron in increased amounts from a few high status graves after 300 BC as sets this time apart from the Mumun pottery period. It is thus that, as a cultural-technical period, the Mumun was finished by approximately 300 BC.

From about 300 BC, bronze objects became the most valued prestige mortuary goods, but iron objects were traded and then produced in the Korean peninsula at that time. The Late Mumun-Early Iron age Neuk-do Island Shellmidden Site yielded a small number of iron objects, Lelang and Yayoi pottery, and other evidence showing that beginning in the Late Mumun, local societies were drawn into closer economic and political contact with the societies of the Late Zhou Dynasty, Final Jōmon, and Early Yayoi.

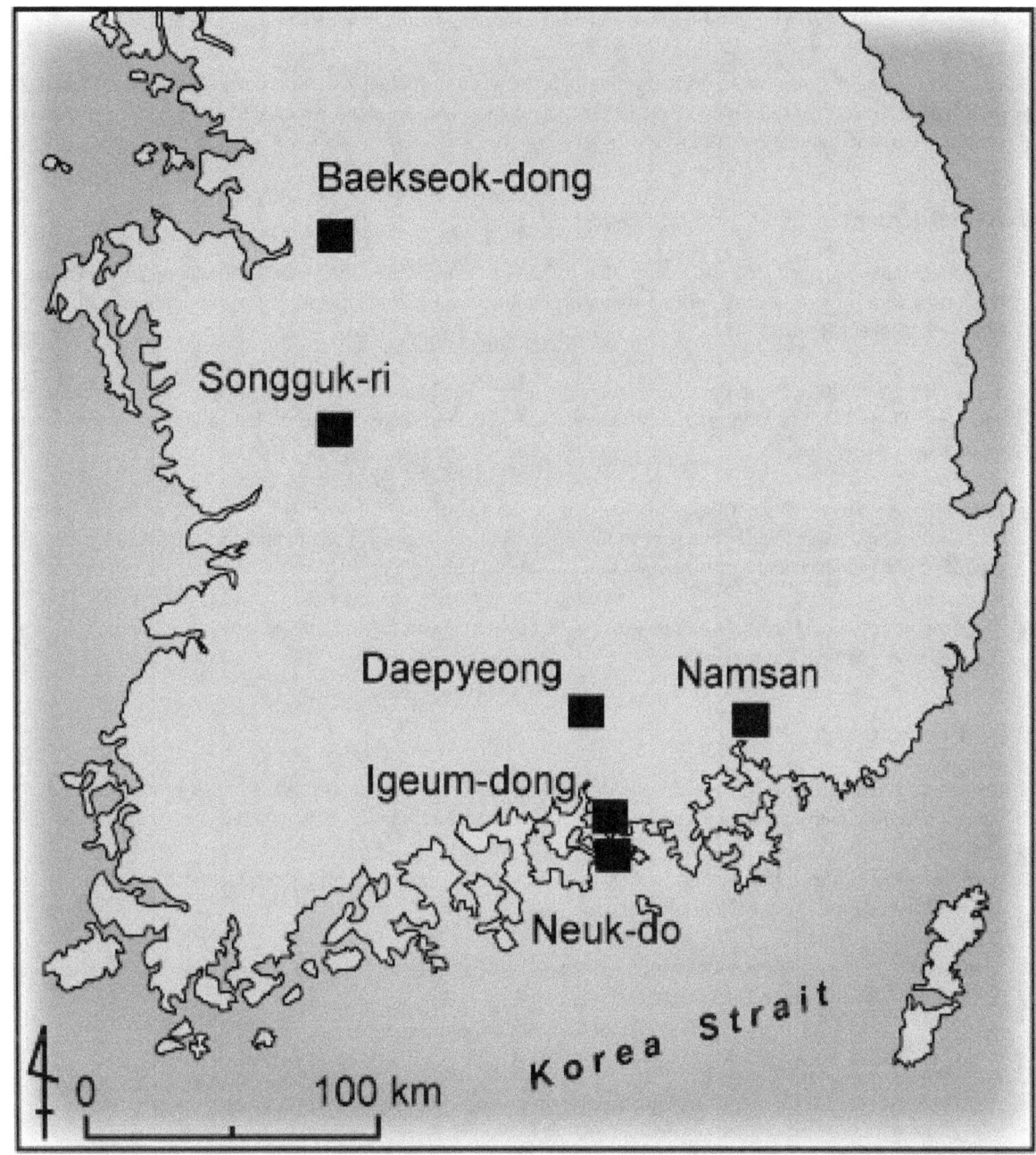

Settlement sites of the Mumun Period that are mentioned in the text of this article.

10.2 Mumun cultural traits

As an archaeological culture, the Mumun is composed of the following elements:

10.2.1 Subsistence

- Broad-spectrum subsistence was practiced through the Early Mumun. That is to say, evidence excavated from pit-houses and other outdoor household features indicates that hunting, fishing, and foraging was occurring in addition to agriculture.*[6]

- stone tools used in agricultural subsistence activities are common and include semi-lunar blades.*[4]

- Intensive wet-field agriculture (paddy farming) was in place in the Middle Mumun.*[3] However, even the pit-houses of settlements associated with wet-field archaeological features show evidence that people were also engaged to some degree in hunting and fishing.

10.2.2   Settlement

- Large rectangular-shaped pit-houses were used in Early Mumun. These pit-houses had one or more hearths, and pit-houses with up to 6 hearths indicate that such features were the living spaces for multiple generations of the same household.*[8]

- Some time after 900 BC, small pit-houses were the norm. The plan-shape of these pit-houses are square, circular and oval. They do not have interior hearths —instead, the central area of the pit-house floor is equipped with a shallow oval 'work-pit'.*[1]

- Archaeologists see this change in architecture as a social shift in the household. Namely, the tight and multi-generational unit housed under one roof in the Early Mumun changed fundamentally into households formed of groups of semi-independent nuclear family units in separate pit-houses.*[8]

- The average settlement in the Mumun was small, but settlements with as many as several hundred pit-houses emerged in the Middle Mumun.*[8]

10.2.3   Economy

- Household production was the basic mode of the Mumun economy, but specialized craft production and a big-man-style redistributive prestige economy emerged in the Middle Mumun.*[8]

- Archaeological evidence has documented cases in which it appears that surplus production of crops, stone tools, and pottery occurred in the Middle Mumun.*[3]*[4]

- Artifacts that illustrate regional redistributive systems and exchange include greenstone ornaments, bronze objects, and some kinds of red-burnished pottery.*[9]

10.2.4   Mortuary practices

- Megalithic burials, stone-cist burials, and jar burials are found.

- Some burials in the latter part of the Middle Mumun are especially large and required a significant amount of labour to construct. A small number of Middle Mumun burials contain prestige/ceremonial artifacts such as bronze, greenstone, groundstone daggers, and red-burnished ware.*[4]*[9]*[10]

10.3   See also

- List of archaeological periods - master list

- List of Korea-related topics

- Prehistoric Korea

- Liaoning bronze dagger culture

10.4 References

[1] Ahn, Jae-ho (2000). "Hanguk Nonggyeongsahoe-eui Seongnib (The Formation of Agricultural Society in Korea)". *Hanguk Kogo-Hakbo* (in Korean) **43**: 41–66.

[2] Bale, Martin T. (2001). "Archaeology of Early Agriculture in Korea: An Update on Recent Developments". *Bulletin of the Indo-Pacific Prehistory Association* **21** (5): 77–84.

[3] Crawford, Gary W.; Gyoung-Ah Lee (2003). "Agricultural Origins in the Korean Peninsula". *Antiquity* **77** (295): 87–95.

[4] Rhee, S. N.; Choi, M. L. (1992). "Emergence of Complex Society in Prehistoric Korea". *Journal of World Prehistory* **6**: 51–95. doi:10.1007/BF00997585.

[5] Kim, Seung Og (1996). *Political Competition and Social Transformation: The Development of Residence, Residential Ward, and Community in Prehistoric Taegongni of Southwestern Korea.* Ann Arbor: University of Michigan Press.

[6] Lee, June-Jeong (2001). *From Shellfish Gathering to Agriculture in Prehistoric Korea: The Chulmun to Mumun Transition.* Madison: University of Wisconsin-Madison Press.

[7] Kim, Jangsuk (2003). "Land-use Conflict and the Rate of Transition to Agricultural Economy: A Comparative Study of Southern Scandinavia and Central-western Korea". *Journal of Archaeological Method and Theory* **10** (3): 277–321. doi:10.1023/A:1026087723164.

[8] Bale, Martin T.; Min-jung Ko (2006). "Craft Production and Social Change in Mumun Period Korea". *Asian Perspectives* **45** (2): 159–187. doi:10.1353/asi.2006.0019.

[9] Nelson, Sarah M. (1999). "Megalithic Monuments and the Introduction of Rice into Korea". In Gosden, C.; Hather, J. *The Prehistory of Food: Appetites for Change.* London: Routledge. pp. 147–165. ISBN 0-415-11765-8.

[10] Bale, Martin T. "Excavations of Large-scale Megalithic Burials at Yulha-ri, Gimhae-si, Gyeongsang Nam-do". Korea Institute, Harvard University. Retrieved 2007-11-08.

10.5 Further reading

• Nelson, Sarah M. (1993). *The Archaeology of Korea.* Cambridge: Cambridge University Press. ISBN 0-521-40443-6.

10.6 External links

• Paper on Boseong River excavations, by Kim Gyeongtaek

Chapter 11

Liaoning bronze dagger culture

The **Liaoning bronze dagger culture** is an archeological complex of the late Bronze Age in Korea and China. Artifacts from the culture are found primarily in the Liaoning area of northeast China and in the Korean peninsula. Various other bronze artifacts, including ornaments and weapons, are associated with the culture, but the daggers are viewed as the most characteristic. Liaoning bronzes contain a higher percentage of zinc than those of the neighboring bronze cultures.[*][1]

Lee Chung-kyu (1996) considers that the culture is properly divided into five phases: Phases I and II typified by violin-shaped daggers, Phases IV and V by slender daggers, and Phase III by the transition between the two. Of these, remains from Phases I, II and III can be found in some amounts in both the Korean peninsula and northeast China, but remains from Phases IV and V are found almost exclusively in Korea.

11.1   Violin-shaped daggers

The early phase consists of an early period of bronze manufacture without daggers, followed by a period of producing violin-shaped daggers. The prime period of production of violin-shaped daggers is dated to the 8th and 7th centuries BCE.

The earliest artifacts from this period are found exclusively in China (mostly in the former territory of Gojoseon - the territory of the kingdom is approximated by the distribution of violin-shaped daggers and table-shaped dolmens.), and seem only gradually to have spread to the Korean peninsula. By Lee's (1996) Phase II, however, a distinctive notched form of dagger begins to emerge in southern Korea, suggesting that by this time independent bronze production had begun in that region.

Evidence gained from pottery indicates that the bronze dagger "culture" of this time actually included several distinct cultural groups. One distinct pottery tradition is found in northeast China and northwestern Korea, another in the Taedong River valley, another in the southwest around the Chungcheong provinces including the Geum River, and yet another throughout the rest of the southern Korean peninsula including Jeju island.

11.2   Slender daggers

This later part of the Liaoning bronze dagger culture is often referred to as the "Korean bronze dagger culture," since it was largely restricted to the Korean peninsula. At this point the Liaoning culture artifacts begin to disappear from the northeast China area. A new form of dagger begins to turn up on the Korean peninsula, straight and slender.

The greatest concentration of bronze daggers is found in the Geum River valley of South Chungcheong province. Away from this area, the daggers become progressively fewer. This appears to indicate that most daggers were produced in the Geum valley, and the other cultures of the peninsula acquired them primarily by trade. Trade also took place by sea, with artifacts from the Later Phase found in Japanese archeological sites as well.

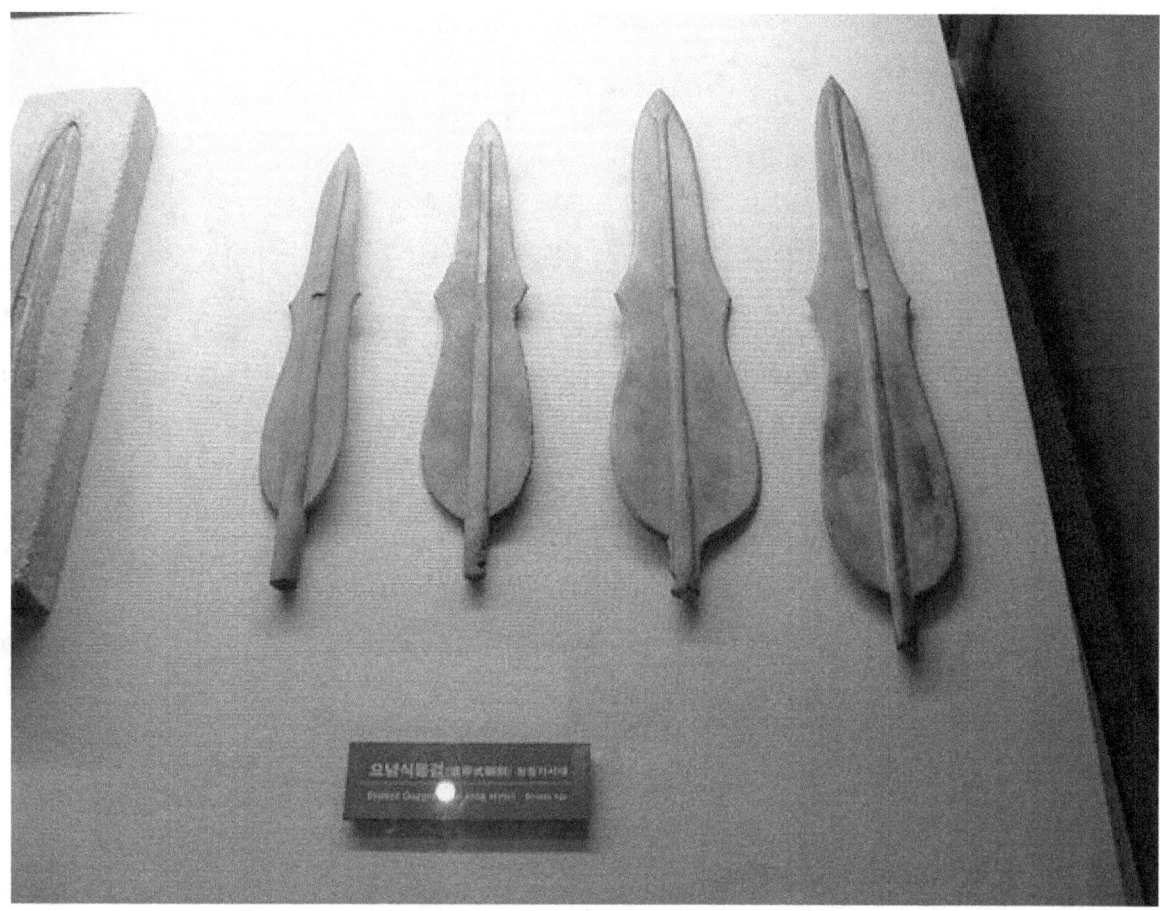

Liaoning-style violin-shaped bronze knives from Korea held at the War Memorial (Seoul).

Lee (1996) divides this phase into two distinct sections: one dating to the 3rd century BCE in which the production of slender bronze daggers predominated, and one dating to the 2nd century BCE in which daggers are often accompanied by bronze mirrors with geometric designs and halberds influenced by the Chinese Qin state. In the first part, a single pottery culture typified by clay-band applique is found throughout the Korean peninsula, but in the second part distinctive pottery types emerge in the northwest and the remainder of the peninsula.

11.3   Historical identity

The disappearance of the Liaoning bronze dagger culture from China appears to coincide with the State of Yan's conquest of that area. The Korean bronze dagger culture of the Later Phase appears to correspond with the state of Jin, which occasionally enters Chinese annals as a contemporary of Wiman Joseon. Lee (1984, p. 13) views this as the period of emergence of the "walled-town states" in Korean culture, a hierarchical political structure in contrast to the tribal system which had prevailed during the Neolithic period.

11.4   See also

- List of China-related topics

- List of Korea-related topics

- History of Korea

Chapter 12

Bronze Age Europe

The **European Bronze Age** is characterized by bronze artifacts and the use of bronze implements. The regional Bronze Age succeeds the Neolithic. It starts with the Aegean Bronze Age in 3200 BCE*[1] (succeeded by the Beaker culture), and spans the entire 2nd millennium BCE (Unetice culture, Tumulus culture, Terramare culture, Urnfield culture and Lusatian culture) in Northern Europe, lasting until c. 600 BCE.

12.1 History

12.1.1 Aegean

The Aegean Bronze Age begins around 3200 BCE*[1] when civilizations first established a far-ranging trade network. This network imported tin and charcoal to Cyprus, where copper was mined and alloyed with the tin to produce bronze. Bronze objects were then exported far and wide, and supported the trade. Isotopic analysis of the tin in some Mediterranean bronze objects indicates it came from as far away as Great Britain.

Knowledge of navigation was well developed at this time, and reached a peak of skill not exceeded until a method was discovered (or perhaps rediscovered) to determine longitude around AD 1750, with the notable exception of the Polynesian sailors.

The Minoan civilization based from Knossos appears to have coordinated and defended its Bronze Age trade. One crucial lack in this period was that modern methods of accounting were not available.

12.1.2 Caucasus

The Maykop culture was the major early Bronze Age culture in the North Caucasus. Some scholars date arsenical bronze artifacts in the region as far back as the mid-4th millennium BCE.*[2]

12.1.3 Eastern Europe

Main article: Aegean Bronze Age
See also: Bronze Age Southeastern Europe and Bronze Age Romania

The Yamna culture*[3] is a late copper age/early Bronze Age culture dating to the 36th–23rd centuries BCE. The culture was predominantly nomadic, with some agriculture practiced near rivers and a few hill-forts.

The Catacomb culture, covering several related archaeological cultures, was first to introduce corded pottery decorations into the steppes and showed a profuse use of the polished battle axe, providing a link to the West. Parallels with the

A display of Late Bronze Age vessels and tools from various Romanian locations, at the National Museum of the Union, Alba Iulia

Afanasevo culture, including provoked cranial deformations, provide a link to the East. It was preceded by the Yamna culture and succeeded by the western Corded Ware culture. The Catacomb culture in the Pontic steppe was succeeded by the Srubna culture from c. the 17th century BCE.

12.1.4   Central Europe

See also: Bronze Age Transylvania and Pre-Celtic

Important sites include:

- Biskupin (Poland)

- Nebra (Germany)

- Zug-Sumpf, Zug, Switzerland

- Vráble, Slovakia

In Central Europe, the early Bronze Age Unetice culture (1800-1600 BCE) includes numerous smaller groups like the Straubingen, Adlerberg and Hatvan cultures. Some very rich burials, such as the one located at Leubingen (today part of Sömmerda) with grave gifts crafted from gold, point to an increase of social stratification already present in the Unetice culture. All in all, cemeteries of this period are rare and of small size. The Unetice culture is followed by the middle

Bronze Age (1600-1200 BC) Tumulus culture, which is characterised by inhumation burials in tumuli (barrows). In the eastern Hungarian Körös tributaries, the early Bronze Age first saw the introduction of the Makó culture, followed by the Otomani and Gyulavarsánd cultures.

The late Bronze Age Urnfield culture, (1300 BCE−700 BCE) is characterized by cremation burials. It includes the Lusatian culture in eastern Germany and Poland (1300-500 BCE) that continues into the Iron Age. The Central European Bronze Age is followed by the Iron Age Hallstatt culture (700-450 BCE).

12.1.5 Northern Europe

Main article: Nordic Bronze Age

In northern Germany, Denmark, Sweden and Norway, Bronze Age inhabitants manufactured many distinctive and artistic artifacts, such as the pairs of lurer horns discovered in Denmark. Some linguists believe that an early Indo-European language was introduced to the area probably around 2000 BCE, which eventually became Proto-Germanic, the last common ancestor of the Germanic languages. This would fit with the apparently unbroken evolution of the Nordic Bronze Age into the most probably ethnolinguistically Germanic Pre-Roman Iron Age.

The age is divided into the periods I-VI, according to Oscar Montelius. Period Montelius V, already belongs to the Iron Age in other regions.

12.1.6 Britain

Main article: Bronze Age Britain

In Great Britain, the Bronze Age is considered to have been the period from around 2100 to 700 BCE. Immigration brought new people to the islands from the continent. Recent tooth enamel isotope research on bodies found in early Bronze Age graves around Stonehenge indicate that at least some of the immigrants came from the area of modern Switzerland. The Beaker people displayed different behaviours from the earlier Neolithic people and cultural change was significant. Integration is thought to have been peaceful as many of the early henge sites were seemingly adopted by the newcomers. The rich Wessex culture developed in southern Britain at this time. Additionally, the climate was deteriorating; where once the weather was warm and dry it became much wetter as the Bronze Age continued, forcing the population away from easily defended sites in the hills and into the fertile valleys. Large livestock ranches developed in the lowlands which appear to have contributed to economic growth and inspired increasing forest clearances. The Deverel-Rimbury culture began to emerge in the second half of the 'Middle Bronze Age' (c. 1400-1100 BCE) to exploit these conditions. Cornwall was a major source of tin for much of western Europe and copper was extracted from sites such as the Great Orme mine in northern Wales. Social groups appear to have been tribal but with growing complexity and hierarchies becoming apparent.

Also, the burial of dead (which until this period had usually been communal) became more individual. For example, whereas in the Neolithic a large chambered cairn or long barrow was used to house the dead, the 'Early Bronze Age' saw people buried in individual barrows (also commonly known and marked on modern British Ordnance Survey maps as Tumuli), or sometimes in cists covered with cairns.

The greatest quantities of bronze objects found in England were discovered in East Cambridgeshire, where the most important finds were done in Isleham (more than 6500 pieces).*[4]

See also: Atlantic Bronze Age

12.1.7 Bronze Age boats

See also: Atlantic Bronze Age

- Ferriby Boats

- Langdon Bay hoard - see also Dover Museum

- Divers unearth Bronze Age hoard off the coast of Devon

- Moor Sands finds, including a remarkably well preserved and complete sword which has parallels with material from the Seine basin of northern France

12.1.8 Ireland

See also: Atlantic Bronze Age

The Bronze Age in Ireland commenced in the centuries around 2000 BCE when copper was alloyed with tin and used to manufacture Ballybeg type flat axes and associated metalwork. The preceding period is known as the Copper Age and is characterised by the production of flat axes, daggers, halberds and awls in copper. The period is divided into three phases: Early Bronze Age 2000-1500 BCE; Middle Bronze Age 1500-1200 BCE and Late Bronze Age 1200-c.500 BCE. Ireland is also known for a relatively large number of Early Bronze Age Burials.,[5] [6]

12.2 See also

- Chariot burial

- Megalithic tomb

- Old European hydronymy

- Helladic period

- Nordic Bronze Age

- Atlantic Bronze Age

12.3 References

[1] "Ancient Greece". British Museum. Retrieved 2015-05-06.

[2] Douglas Q. Adams (January 1997). *Encyclopedia of Indo-European Culture*. Taylor & Francis. pp. 372–374. ISBN 978-1-884964-98-5.

[3] Also known as Pit Grave culture or Ochre Grave culture

[4] Hall, David (1994). *Fenland survey : an essay in landscape and persistence / David Hall and John Coles.* London; English Heritage. ISBN 1-85074-477-7., p. 81-88

[5] Waddell, J. 1998. *The Prehistoric Archaeology of Ireland.* Galway.

[6] Eogan, G. 1983. *The Hoards of the Irish Later Bronze Age.* Dublin

Chapter 13

Late Bronze Age collapse

The fall of Troy, an event recounted in Greek mythology at the end of the Bronze Age, as represented by the 17th century painter Kerstiaen De Keuninck

The **Late Bronze Age collapse** was a transition in the Aegean Region, Southwestern Asia and the Eastern Mediterranean from the Late Bronze Age to the Early Iron Age that historians believe was violent, sudden and culturally disruptive. The palace economy of the Aegean Region and Anatolia which characterised the Late Bronze Age was replaced, after a hiatus, by the isolated village cultures of the Greek Dark Ages.

Between 1206 and 1150 BC, the cultural collapse of the Mycenaean kingdoms, the Hittite Empire in Anatolia and Syria,*[1] and the New Kingdom of Egypt in Syria and Canaan*[2] interrupted trade routes and severely reduced literacy. In the first phase of this period, almost every city between Pylos and Gaza was violently destroyed, and often left unoccupied thereafter: examples include Hattusa, Mycenae, and Ugarit.*[3] Drews writes "Within a period of forty to fifty years at the end of the thirteenth and the beginning of the twelfth century almost every significant city in the eastern

Mediterranean world was destroyed, many of them never to be occupied again" (p. 4).

The gradual end of the Dark Age that ensued saw the eventual rise of settled Syro-Hittite states in Cilicia and Syria, Aramaean kingdoms of the mid-10th century BC in the Levant, the eventual rise of the Neo-Assyrian Empire, and after the Orientalising period of the Aegean, Classical Greece.

13.1 Regional evidence

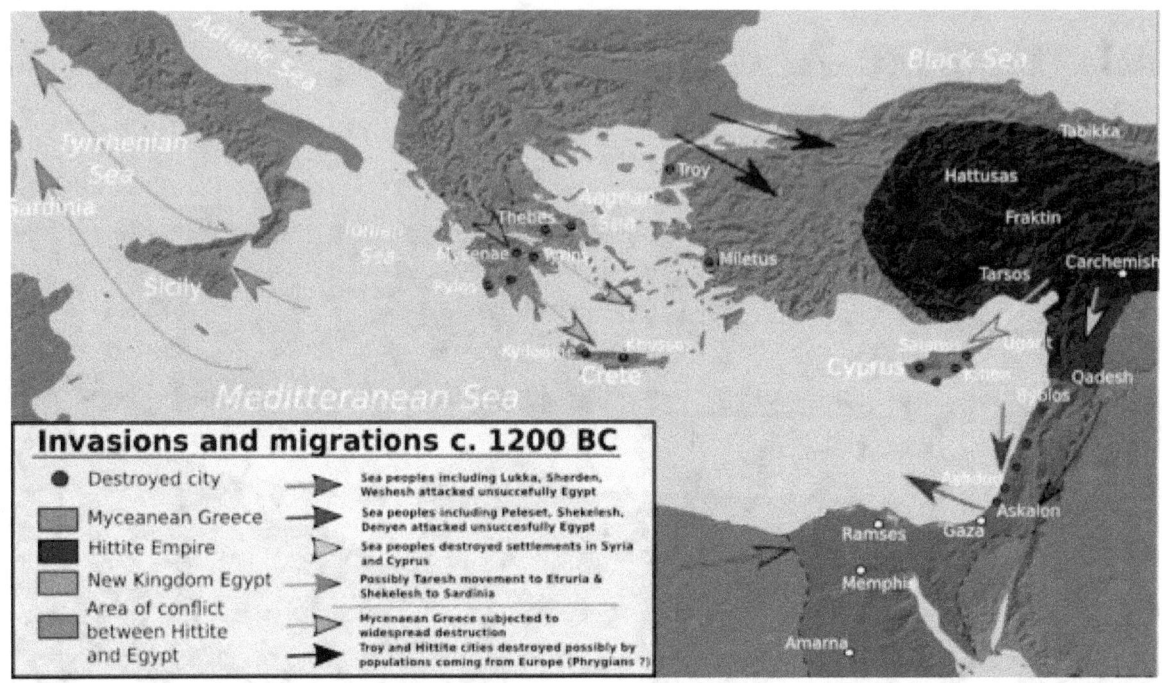

Invasions, destructions and possible population movements during the collapse of the Bronze Age, ca. 1200.

13.1.1 Evidence of destruction

Anatolia

Prior to the Bronze Age collapse, Anatolia (Asia Minor) was dominated by a number of Indo-European peoples: Luwians, Hittites, Mitanni, and Mycenaean Greeks, together with the Semitic Assyrians. From the 17th century BC, the Mitanni formed a ruling class over the Hurrians, an ancient indigenous Caucasian people who spoke a Hurro-Urartian language isolate. Similarly, the Hittites absorbed the Hattians,[4] a people speaking a language which may have been of the North Caucasian group.

Every Anatolian site that was important during the preceding Late Bronze Age shows a destruction layer, and it appears that here civilization did not recover to the level of the Indo-European Hittites for another thousand years. Hattusas, the Hittite capital, was burned—probably by Kaskians, possibly aided by the Phrygians—abandoned, and never reoccupied. Karaoğlan was burned and the corpses left unburied. The Hittite Empire was destroyed by the Indo-European speaking Phrygians and by the Semitic speaking Aramaeans. Troy was destroyed at least twice, before being abandoned until Roman times.

The Phrygians had arrived probably over the Bosphorus in the 13th century BC, and laid waste to the Hittite Empire (already weakened by defeat at the hands of Kaska[5]), before being checked by the Assyrians in the Early Iron Age of the 9th century BC. Other groups of Indo-European warriors followed into the region, most prominently the Armenians, and even later, by the Cimmerians, and Scythians. The Semitic Arameans, Kartvelian speaking Colchians, and Hurro-Urartuans also made an appearance in parts of the region. Sites in Anatolia showing evidence of the collapse are:

- Troy

- Miletus

- Hattusas[6]

- Mersin

- Tarhuntassa

Cyprus

The catastrophe separates Late Cypriot II (LCII) from the LCIII period, with the sacking and burning of Enkomi, Kition, and Sinda, which may have occurred twice before those sites were abandoned.[7] During the reign of the Hittite king Tudhaliya IV (reigned ca. 1237–1209 BC), the island was briefly invaded by the Hittites,[8] either to secure the copper resource or as a way of preventing piracy. Shortly afterwards, the island was reconquered by his son around 1200 BC. Some towns (Enkomi, Kition, Palaeokastro and Sinda) show traces of destruction at the end of LC IIC. Whether or not this is really an indication of a Mycenean invasion is contested. Originally, two waves of destruction, ca. 1230 BC by the Sea Peoples and ca. 1190 BC by Aegean refugees have been proposed.[9] The smaller settlements of Ayios Dhimitrios and Kokkinokremnos, as well as a number of other sites, were abandoned, but do not show traces of destruction. Kokkinokremos was a short-lived settlement, where various caches concealed by smiths have been found. That no one ever returned to reclaim the treasures suggests that they were killed or enslaved. Recovery only occurred in the Early Iron Age with Phoenician and Greek settlement. Sites in Cyprus showing evidence of the collapse are:

- Palaeokastro

- Kition

- Sinda

- Enkomi

Syria

Ancient Syria had been initially dominated by a number of indigenous Semitic speaking peoples; the Canaanites, Amorites, and cities of Ebla and Ugarit were prominent among these.

Prior to and during the Bronze Age Collapse, Syria became a battle ground between the empires of the Hittites, Assyrians, Mitanni and Egyptians, and the coastal regions came under attack from the Sea Peoples. From the 13th Century BC, the Arameans came to prominence in Syria, and the region outside of the Phoenician coastal areas eventually became Aramaic speaking.

Syrian sites previously showed evidence of trade links with Mesopotamia (Assyria and Babylonia), Egypt and the Aegean in the Late Bronze Age. Evidence at Ugarit shows that the destruction there occurred after the reign of Merneptah (ruled 1213–1203 BC) and even the fall of Chancellor Bay (died 1192 BC). The last Bronze Age king of the Semitic state of Ugarit, Ammurapi, was a contemporary of the Hittite king Suppiluliuma II. The exact dates of his reign are unknown. However, a letter by the king is preserved on one of the clay tablets found baked in the conflagration of the destruction of the city. Ammurapi stresses the seriousness of the crisis faced by many Levantine states from invasion by the advancing Sea Peoples in a dramatic response to a plea for assistance from the king of Alasiya. Ammurapi highlights the desperate situation Ugarit faced in letter RS 18.147:

> My father, behold, the enemy's ships came (here); my cities(?) were burned, and they did evil things in my country. Does not my father know that all my troops and chariots(?) are in the Land of Hatti, and all

my ships are in the Land of Lukka?...Thus, the country is abandoned to itself. May my father know it: the seven ships of the enemy that came here inflicted much damage upon us.*[10]

Unfortunately for Ugarit, no help arrived and Ugarit was burned to the ground at the end of the Bronze Age. Its destruction levels contained Late Helladic IIIB ware, but no LH IIIC (see Mycenaean period). Therefore, the date of the destruction is important for the dating of the LH IIIC phase. Since an Egyptian sword bearing the name of pharaoh Merneptah was found in the destruction levels, 1190 BC was taken as the date for the beginning of the LH IIIC. A cuneiform tablet found in 1986 shows that Ugarit was destroyed after the death of Merneptah. It is generally agreed that Ugarit had already been destroyed by the 8th year of Ramesses III—i e. 1178 BC. These letters on clay tablets found baked in the conflagration of the destruction of the city speak of attack from the sea, and a letter from Alashiya (Cyprus) speaks of cities already being destroyed from attackers who came by sea. It also speaks of the Ugarit fleet being absent, patrolling the Lycian coast.

The West Semitic Arameans eventually superseded the earlier Amorites, Canaanites and people of Ugarit, to whom they were ethno-linguistically related. The Arameans came to dominate the region both politically and militarily from the mid 11th century BC until the rise of the Neo Assyrian Empire in the late 10th century BC, after which the entire region fell to Assyria. Sites in Syria showing evidence of the collapse are:

- Ugarit
- Tell Sukas
- Kadesh
- Qatna
- Hamath
- Alalakh
- Aleppo
- Carchemish
- Emar

Southern Levant

Egyptian evidence shows that, from the reign of Horemheb (ruled either 1319 or 1306 to 1292 BC), wandering Shasu were more problematic than the earlier Apiru. Ramesses II (ruled 1279–1213 BC) campaigned against them, pursuing them as far as Moab, where he established a fortress, after the near collapse at the Battle of Kadesh. During the reign of Merneptah, the Shasu threatened the "Way of Horus" north from Gaza. Evidence shows that Deir Alla (Succoth) was destroyed after the reign of Queen Twosret (ruled 1191–1189 BC). The destroyed site of Lachish was briefly reoccupied by squatters and an Egyptian garrison, during the reign of Ramesses III (ruled 1186–1155 BC). All centres along a coastal route from Gaza northward were destroyed, and evidence shows Gaza, Ashdod, Ashkelon, Akko, and Jaffa were burned and not reoccupied for up to thirty years. Inland Hazor, Bethel, Beit Shemesh, Eglon, Debir, and other sites were destroyed. Refugees escaping the collapse of coastal centres may have fused with incoming nomadic and Anatolian elements to begin the growth of terraced hillside hamlets in the highlands region that was associated with the later development of the Hebrews.*[11] During the reign of Rameses III Philistines were allowed to resettle the coastal strip from Gaza to Joppa, Denyen (possibly the tribe of Dan in the Bible, or more likely the people of Adana, also known as Danuna, part of the Hittite Empire) settled from Joppa to Acre, and Tjekker in Acre. These sites quickly achieved independence as the Tale of Wenamun shows. Sites in Southern Levant showing evidence of the collapse are:

- Hazor
- Akko
- Megiddo

- Deir ´Alla

- Bethel

- Beth Shemesh

- Lachish

- Ashod

- Ashkelon

Greece

Main article: Greek Dark Ages

None of the Mycenaean palaces of the Late Bronze Age survived (with the possible exception of the Cyclopean fortifications on the Acropolis of Athens) with destruction being heaviest at palaces and fortified sites. Up to 90% of small sites in the Peloponnese were abandoned, suggesting a major depopulation. The End Bronze Age collapse marked the start of what has been called the Greek Dark Ages, which lasted for more than 400 years. Other cities, like Athens, continued to be occupied, but with a more local sphere of influence, limited evidence of trade and an impoverished culture, from which it took centuries to recover. Sites in Greece showing evidence of the collapse are:

- Teichos Dymaion

- Pylos

- Nichoria

- The Menelaion

- Tiryns

- Mycenae

- Thebes

- Lefkandi

- Iolkos[*][12]

- Knossos

- Kydonia

13.1.2 Areas which marginally survived

Mesopotamia

The Middle Assyrian Empire controlled colonies in Anatolia, which came under attack from the Mushki. Tiglath-Pileser I (reigned 1114–1076 BC) was able to defeat and repel these attacks. The Assyrian Empire survived intact throughout much of this period, with Assyria dominating and often ruling Babylonia directly, controlling south east and south western Anatolia, north western Iran and much of northern and central Syria and Canaan, as far as the Mediterranean and Cyprus. The Arameans and Phrygians were subjected, and Assyria and its colonies were not threatened by the Sea Peoples. However, after the death of Tiglath-Pileser I in 1076 BC, Assyria withdrew to its natural borders in northern Mesopotamia. Assyria retained a stable monarchy, the best army in the world and an efficient civil administration, thus enabling it to survive the Bronze Age Collapse intact and, from the late 10th Century BC, it once more began to assert itself internationally.*[13] However, the situation in Babylonia was very different: after the Assyrian withdrawal, new groups of Semites, such as the Aramaeans and later Chaldeans and Suteans, spread unchecked into Babylonia, and the control by its weak kings barely extended beyond the city limits of Babylon. Babylon was sacked by the Elamites under Shutruk-Nahhunte (ca. 1185–1155 BC), and lost control of the Diyala River valley to Assyria.

Egypt

Main article: Third Intermediate Period of Egypt

After apparently surviving for a while, the Egyptian Empire collapsed in the mid twelfth century BC (during the reign of Ramesses VI, 1145 to 1137 BC). Previously, the Merneptah Stele (ca. 1200 BC) spoke of attacks from Libyans, with associated people of Ekwesh, Shekelesh, Lukka, Shardana and Tursha or Teresh possibly Troas, and a Canaanite revolt, in the cities of Ashkelon, Yenoam and the people of Israel. A second attack during the reign of Ramesses III (1186–1155 BC) involved Peleset, Tjeker, Shardana and Denyen.

13.1.3   Conclusion

Robert Drews describes the collapse as "the worst disaster in ancient history, even more calamitous than the collapse of the Western Roman Empire".*[14] A number of people have spoken of the cultural memories of the disaster as stories of a "lost golden age". Hesiod for example spoke of Ages of Gold, Silver and Bronze, separated from the modern harsh cruel world of the Age of Iron by the Age of Heroes. Rodney Castledon even suggests that memories of the Bronze Age collapse even coloured Plato's story of Atlantis*[15] in the Timaeus and the Critias.

13.2   Possible causes of collapse

There are various theories put forward to explain the situation of collapse, many of them compatible with each other.

13.2.1   Environmental

Climate change

Main article: Bond event

Changes in climate similar to the Younger Dryas period or the Little Ice Age punctuate human history. The local effects of these changes may cause crop failures in multiple consecutive years, leading to warfare as a last-ditch effort at survival. The triggers for climate change are still debated, but ancient peoples could neither have predicted nor coped with substantial climate changes.

Volcanoes

The Hekla 3 eruption approximately coincides with this period and, while the exact date is under considerable dispute, one group calculated the date specifically to be 1159 BC and implicated the eruption in the collapse in Egypt.*[16]

Drought

Using the Palmer Drought Index for 35 Greek, Turkish, and Middle Eastern weather stations, it was shown that a drought of the kind that persisted from January 1972 would have affected all of the sites associated with the Late Bronze Age collapse.*[17]*[18] Drought could have easily precipitated or hastened socio-economic problems and led to wars. More recently it has been shown how the diversion of mid-winter storms, from the Atlantic to north of the Pyrenees and the Alps, bringing wetter conditions to Central Europe but drought to the Eastern Mediterranean, was associated with the Late Bronze Age collapse.[19] Pollen in sediment cores from the Dead Sea and the Sea of Galilee show that there was a period of severe drought at the start of the collapse.[20][21]

13.2.2 Cultural

Migrations and raids

Further information: Sea peoples, Dorian Invasion, Mushki, Aramaeans and Ancient Iranian peoples

Ekrem Akurgal, Gustav Lehmann and Fritz Schachermeyer—following the views of Gaston Maspero—have argued for this view.

Evidence includes the widespread findings of *Naue II-type swords* (coming from South-Eastern Europe) throughout the region, and Egyptian records of invading "northerners from all the lands".

The Ugarit correspondence at the time mentions invasions by tribes of the mysterious Sea Peoples, who appear to have been a disparate mix of Luwians, Greeks and Canaanites, among others. Equally, the last Greek Linear B documents in the Aegean (dating to just before the collapse) reported a large rise in piracy, slave raiding and other attacks, particularly around Anatolia. Later fortresses along the Libyan coast, constructed and maintained by the Egyptians after the reign of Ramesses II, were built to reduce raiding.

This theory is strengthened by the fact that the collapse coincides with the appearance in the region of many new ethnic groups. These include Indo-European tribes, such as the Phrygians, Proto-Armenians, Medes, Persians, Cimmerians, Lydians and Scythians, as well as the Pontic-speaking Colchians, Hurro-Urartuans and Iranian Sarmatians. These groups settled or emerged in the Caucasus, Iran and Anatolia. Thracians, Macedonians and Dorian Greeks seem to have arrived at this time—possibly from the north, usurping the earlier Greeks of Mycenae and Achaea. There also seems to have been widespread migration of Semitic peoples, such as Aramaeans, Chaldeans and Suteans—possibly from the South-East.

The ultimate reasons for these migrations could include drought, developments in warfare/weaponry, earthquakes, or other natural disasters, meaning that the Migrations theory is not necessarily incompatible with the other theories mentioned here. Manuel Robbins for example writes "There is no doubt that people, 'barbarians' or otherwise, were on the move, and some were probably responsible for disruption and attacks on cities. But it is reasonable to believe that they were victims of circumstances themselves and not the initial cause or main agent of disruption." [22]

Ironworking

The Bronze Age collapse may be seen in the context of a technological history that saw the slow, comparatively continuous spread of iron-working technology in the region, beginning with precocious iron-working in what is now Bulgaria and Romania in the 13th and 12th centuries BC.[23]

Leonard R. Palmer suggested that iron, while inferior to bronze weapons, was in more plentiful supply and so allowed larger armies of iron users to overwhelm the smaller armies of bronze-using maryannu chariotry.[24] This argument has been weakened of late with the finding that the shift to iron occurred *after* the collapse, not before. It now seems that the disruption of long distance trade—an aspect of "systems collapse"—cut easy supplies of tin, making bronze impossible to make. Older implements were recycled and then iron substitutes were used.

Changes in warfare

Robert Drews argues[25] that the appearance of massed infantry, using newly developed weapons and armor, such as cast rather than forged spearheads and long swords, a revolutionizing cut-and-thrust weapon,[26] and javelins, and the appearance of bronze foundries, suggest "that mass production of bronze artifacts was suddenly important in the Aegean". (For example, Homer uses "spears" as a virtual synonym for "warriors".) Such new weaponry, in the hands of large numbers of "running skirmishers" who could swarm and cut down a chariot army, would destabilize states that were based upon the use of chariots by the ruling class and precipitate an abrupt social collapse as raiders began to conquer, loot, and burn the cities.[27][28][29]

13.2.3 General systems collapse

Main article: Societal collapse

A general systems collapse has been put forward as an explanation for the reversals in culture that occurred between the Urnfield culture of the 12–13th centuries BC and the rise of the Celtic Hallstatt culture in the 9th and 10th centuries BC.[30] This theory may, however, simply raise the question of whether this collapse was the cause of, or the effect of, the Bronze Age collapse being discussed. General Systems Collapse theory, pioneered by Joseph Tainter,[31] hypothesizes how social declines in response to complexity may lead to a collapse resulting in simpler forms of society.

In the specific context of the Middle East, a variety of factors—including population growth, soil degradation, drought, cast bronze weapon and iron production technologies—could have combined to push the relative price of weaponry (compared to arable land) to a level unsustainable for traditional warrior aristocracies. In complex societies that were increasingly fragile and less resilient, this combination of factors may have contributed to the collapse.

The growing complexity and specialization of the Late Bronze Age political, economic, and social organization in Carol Thomas and Craig Conant's phrase[32] is a weakness that could explain such a widespread collapse that was able to render the Bronze Age civilizations incapable of recovery. The critical flaws of the Late Bronze Age are its centralization, specialization, complexity and top-heavy political structure. These flaws then revealed themselves through socio-political factors (revolt of peasantry and defection of mercenaries), fragility of all kingdoms (Mycenaean, Hittite, Ugaritic and Egyptian), demographic crises (overpopulation), and wars between states. Other factors that could have placed increasing pressure on the fragile kingdoms include piratical disturbances of maritime trade by the Sea Peoples, drought, crop failures, famine, Dorian migration or invasion.

13.3   Links to religious and historical texts

The collapse occurred around the time many Jewish and Christian scholars place the character of Moses.[33] The book of Exodus describes the building of "store cities of Raamses, (Pi Rameses) and Pithom (Pi-Atum), built between the reigns of Rameses II and Setnakhte, and a series of events similar to natural disasters and events in Egypt at this time, followed by a return migration of Semitic people from Egypt back into Canaan. In seeming contradiction, the Books of Kings states that construction began on Solomon's temple 480 years after departure from Egypt. This would place the ten plagues and exodus, 300 years earlier, in the middle of the 15th century BC.

The Greek *Iliad*'s historic accuracy is also still disputed, but it is currently accepted that its core events are related to Troy VII also around the time of the collapse. This could very well indicate that the events of the *Iliad* are a later fictionalization of the traumatic events of the Bronze Age collapse and how these events affected the Greeks of the time.

13.4   See also

- Greek Dark Ages—period following the Bronze Age collapse

- Third Intermediate Period of Egypt—a similar period in Egypt

13.5   References

[1] For Syria, see M. Liverani, "The collapse of the Near Eastern regional system at the end of the Bronze Age: the case of Syria" in *Centre and Periphery in the Ancient World*, M. Rowlands, M.T. Larsen, K. Kristiansen, eds. (Cambridge University Press) 1987.

[2] S. Richard, "Archaeological sources for the history of Palestine: The Early Bronze Age: The rise and collapse of urbanism", *The Biblical Archaeologist* (1987)

[3] The physical destruction of palaces and cities is the subject of Robert Drews, *The End of the Bronze Age: changes in warfare and the catastrophe ca. 1200 B.C.*, 1993.

[4] Gurnet, Otto, (1982), *The Hittites* (Penguin) pp. 119–130.

[5] Bryce, Trevor. *The Kingdom of the Hittites.* (Clarendon), p.379

[6] Bryce, Trevor. *The Kingdom of the Hittites* (Clarendon), p. 374.

[7] Robbins, Manuel (2001). *Collapse of the Bronze Age: The Story of Greece, Troy, Israel and Egypt and the Peoples of the Sea.* pp. 220–239

[8] Bryce, Trevor. *The Kingdom of the Hittites* (Clarendon), p. 366.

[9] Paul Aström has proposed dates of 1190 and 1179 BC (Aström).

[10] Jean Nougaryol et al. (1968) Ugaritica V: 87–90 no. 24

[11] Tubbs, Johnathan (1998), "Canaanites" (British Museum Press)

[12] Drews, Robert (1993), *The End of the Bronze Age: Changes in Warfare and the Catastrophe ca. 1200 BC* (Princeton Uni Press)

[13] Georges Roux, *Ancient Iraq*

[14] Drews 1993:1, quotes Fernand Braudel's assessment that the Eastern Mediterranean cultures returned almost to a starting-point ("plan zéro"), "L'Aube", in Braudel, F. (Ed) (1977), *La Mediterranee: l'espace et l'histoire* (Paris)

[15] Castledon, Rodney (1998), "Atlantis Distroyed" (Routledge)]

[16] Yurco, Frank J.. "End of the Late Bronze Age and Other Crisis Periods: A Volcanic Cause". in Teeter, Emily; Larson, John (eds.). *Gold of Praise: Studies on Ancient Egypt in Honor of Edward F. Wente.* (Studies in Ancient Oriental Civilization 58.) Chicago: Oriental Institute of the University of Chicago. 1999:456–458. ISBN 1-885923-09-0.

[17] Weiss, Harvey (June 1982). "The decline of Late Bronze Age civilization as a possible response to climatic change". *Climatic Change* 4 (2): 173–198. doi:10.1007/BF00140587.

[18] Wright, Karen: (1998) "Empires in the Dust" in*Discover*, March 1998. http://discovermagazine.com/1998/mar/empiresinthedus

[19] Fagan, Brian M. (2003). *The Long Summer: How Climate Changed Civilization.* Basic Books.

[20] Kershner, Isabel (22 October 2013). "Pollen Study Points to Drought as Culprit in Bronze Age Mystery". *The New York Times.* doi:10.1006/jasc.1999.0431.

[21] Langgut, Dafna; Finkelstein, Israel ; Litt, Thomas (October 2013) "Climate and the late Bronze Collapse: New evidence from the southern Levant", *Journal of Institute of Archaeology of Tel Aviv University,* **40** (2) : 149–175.

[22] Robbins, Manuel (2001), "Collapse of the Bronze Age: The Story of Greece, Troy, Israel, Egypt and the Pebles of the Sea" (Author's Choice)

[23] See A. Stoia and the other essays in M.L. Stig Sørensen and R. Thomas, eds., *The Bronze Age: Iron Age Transition in Europe* (Oxford) 1989, and T.H. Wertime and J.D. Muhly, *The Coming of the Age of Iron* (New Haven) 1980.

[24] Palmer, Leonard R (1962) *Mycenaeans and Minoans: Aegean Prehistory in the Light of the Linear B Tablets.* (New York, Alfred A. Knopf, 1962)

[25] Drews pp192ff.

[26] The Naue Type II sword, introduced from the eastern Alps and Carpathians ca 1200, quickly established itself and became the only sword in use during the eleventh century; iron was substituted for bronze without essential redesign (Drews 1993:194)

[27] Drews, R. (1993). *The End of the Bronze Age: Changes in Warfare and the Catastrophe ca. 1200 B.C.* (Princeton 1993).

[28]

[29]

[30] http://www.iol.ie/~{}edmo/linktoprehistory.html—a page about the history of Castlemagner, on the web page of the local historical society.

[31] Tainter, Joseph (1976). *The Collapse of Complex Societies* (Cambridge University Press).

[32] Carol G. Thomas and Craig Conant, *Citadel to City-state: The Transformation of Greece, 1200–700 B.C.E.*, 1999.

[33] Robbins, Manuel (2001) op cit, pp240-244

13.6   Further reading

- Dickinson, Oliver (2007). *The Aegean from Bronze Age to Iron Age: Continuity and Change Between the Twelfth and Eighth Centuries BC*. Routledge. ISBN 978-0-415-13590-0.

- Cline, Eric H. (2014). *1177 B.C.: The Year Civilization Collapsed*. Princeton, NJ: Princeton University Press. ISBN 978-0-691-14089-6.

Chapter 14

Prehistory of the Balkans

For the history of Earth before the occupation by the genus homo, *including the period of early hominins, see Geology of Europe and Human evolution.*

See also: Prehistory of Transylvania

The **prehistory of Southeastern Europe** , defined roughly as the territory of the wider Balkans peninsula (including the territories of the modern countries of Albania, Kosovo, Croatia, Serbia, Macedonia, Greece, Bosnia, Hungary, Romania, Bulgaria, Moldova and Turkey) covers the period from the Upper Paleolithic, beginning with the presence of Homo sapiens in the area some 44,000 years ago, until the appearance of the first written records in Classical Antiquity, in Greece as early as the 8th century BC.

Human prehistory in Southeastern Europe is conventionally divided into smaller periods, such as Upper Paleolithic, Holocene Mesolithic/Epipaleolithic, Neolithic Revolution, expansion of Proto-Indo-Europeans, and Protohistory. The changes between these are gradual. For example, depending on interpretation, protohistory might or might not include Bronze Age Greece (2800–1200 BC),[*][1] Minoan, Mycenaean, Thracian, Lemnian, and Venetic cultures. By one interpretation of the historiography criterion, Southeastern Europe enters protohistory only with Homer (*See also Historicity of the Iliad, and Geography of the Odyssey*). At any rate, the period ends before Herodotus in the 5th century BC.[*][2]

14.1   Paleolithic

(2,600,000 – 13,000 BP)

See also: Paleolithic Europe and Paleolithic Transylvania

14.1.1   Balkan Transition to the Upper Paleolithic

(2,600,000 – 50,000 BP BP)

Main article: Kozarnika

The earliest evidence of human occupation discovered in the Balkans, in Kozarnika Bulgaria, date from at least 1.4 million years ago.[*][3]

There is evidence of human presence in the Balkans from the Lower Paleolithic onwards, but the number of sites is limited. According to Douglass W. Bailey:[*][4]

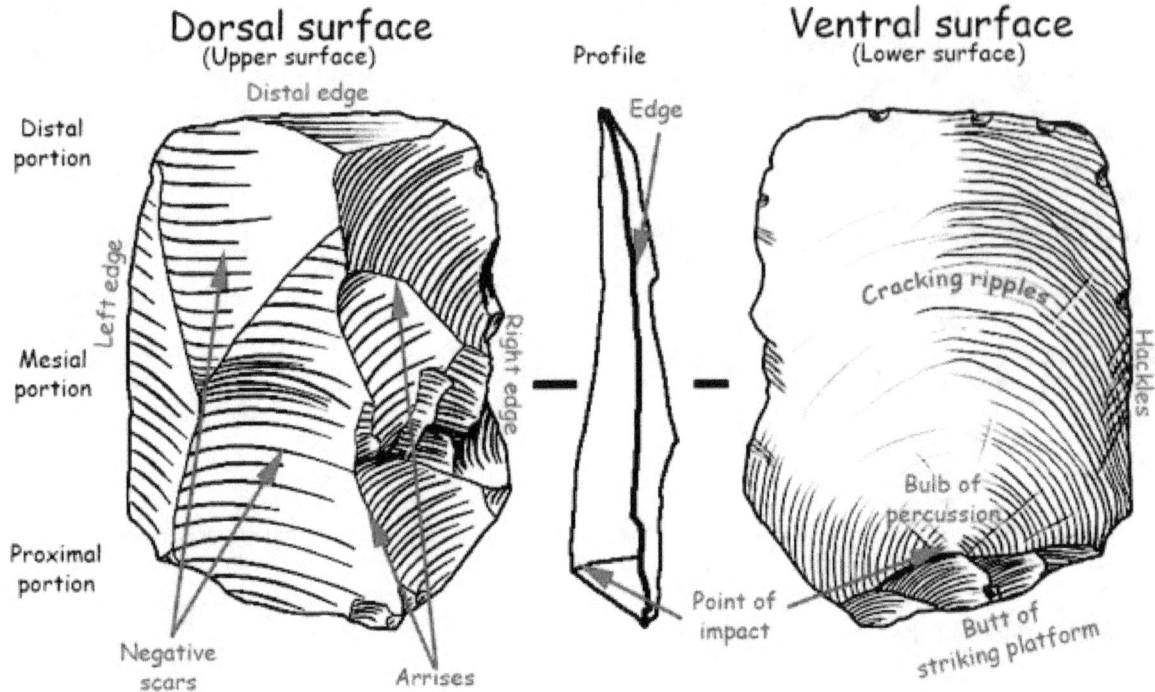

Fundamental elements for the technic description of a lithic flake

The Palaeolithic period, literally the "Old Stone Age", is an ancient cultural level of human development characterized by the use of unpolished chipped stone tools. The transition from Middle to Upper Palaeolithic is directly related to the development of behavioural modernity by homonids around 40,000 years BP. To denote the great significance and degree of change, this dramatic shift from Middle to Upper Palaeolithic is sometimes called the Upper Palaeolithic Revolution.

In the late Pleistocene, various components of the transition–material culture and environmental features (climate, flora, and fauna) indicate continual change, differing from contemporary points in other parts of Europe. The aforementioned aspects leave some doubt that the term Upper Palaeolithic Revolution is appropriate to the Balkans.

In general, continual evolutionary changes are the first crucial characteristic of the transition to the Upper Palaeolithic in the Balkans. The notion of the Upper Palaeolithic Revolution that has been developed for core European regions is not applicable to the Balkans. What is the reason? This particularly significant moment and its origins are defined and enlightened by other characteristics of the transition to upper Old Stone Age. The environment, climate, flora and fauna corroborate the implications.

During the last interglacial period and the most recent glaciation of the Pleistocene (from 131,000 till 12,000 BP), Europe was very different from the Balkans. The glaciations did not affect southeastern Europe to the extent that they did in the northern and central regions. The evidence of forest and steppe indicate the influence was not so drastic; some species of flora and fauna survived only in the Balkans. The Balkans today still abound in species endemic only to this part of Europe.

The notion of gradual transition (or evolution) best defines Balkan Europe from about 50,000 BP. In this sense, the material culture and natural environment of the Balkans of the late Pleistocene and the early Holocene were distinct from other parts of Europe. Douglass W. Bailey writes in *Balkan Prehistory: Exclusion, Incorporation and Identity*: "Less dramatic changes to climate, flora and fauna resulted in less dramatic adaptive, or reactive, developments in material culture."

Thus, in speaking about southeastern Europe, many classic conceptions and systematizations of human development during the Palaeolithic (and then by implication the Mesolithic) should not be considered correct in all cases. In this regard, the absence of Upper Palaeolithic cave art in the Balkans does not seem to be surprising. Civilisations develop new and distinctive characteristics as they respond to new challenges in their environment.

Aurignacian double edged scraper on blade - 3 views of the same object.

14.1.2   Upper Palaeolithic

(50,000 – 20,000 BP)

Main articles: Peștera cu Oase and Peștera Muierilor

In 2002, some of the oldest modern human (Homo sapiens sapiens) remains in Europe were discovered in the "Cave With Bones" (*Peștera cu Oase*), near Anina, Romania.[5] Nicknamed "John of Anina" (*Ion din Anina*), the remains (the lower jaw) are approximately 37,800 years old.

These are some of Europe's oldest remains of *Homo sapiens*, so they are likely to represent the first such people to have entered the continent.[6] According to some researchers, the particular interest of the discovery resides in the fact that it presents a mixture of archaic, early modern human and Neanderthal morphological features,[7] indicating considerable Neanderthal/modern human admixture,[8] which in turn suggests that, upon their arrival in Europe, modern humans met and interbred with Neanderthals. Recent reanalysis of some of these fossils has challenged the view that these remains represent evidence of interbreeding.[9] A second expedition by Erik Trinkaus and Ricardo Rodrigo, discovered further fragments (for example, a skull dated ~36,000, nicknamed "Vasile").

Two human fossil remains found in the Muierii (*Peștera Muierilor*) and the Cioclovina caves in Romania have been radiocarbon dated using the technique of the accelerator mass spectrometry to the age of ~ 30,000 years BP.(see *Human fossil bones from the Muierii Cave and the Cioclovina Cave, Romania*).

The first skull, scapula and tibia remains were found in 1952 in Baia de Fier, in the Muierii Cave, Gorj County in the Oltenia province, by Constantin Nicolaescu-Plopşor.

In 1941 another skull was found at the Cioclovina Cave near Commune Bosorod, Hunedoara County, in Transylvania. The anthropologist, Francisc Rainer, and the geologist, Ion Th. Simionescu, published a study of this skull.

The physical analysis of these fossils was begun in the summer of the year 2000 by Emilian Alexandrescu, archaeologist at the Vasile Pârvan Institute of Archaeology in Bucharest, and Agata Olariu, physicist at the Institute of Physics and Nuclear Engineering-Horia Hulubei, Bucharest, where samples were taken. One sample of bone was taken from the skull from Cioclovina; samples were also taken from the scapula and tibia remains from Muierii Cave. The work continued at the University of Lund, AMS group, by Göran Skog, Kristina Stenström and Ragnar Hellborg. The samples of bones were dated by radiocarbon method applied at the AMS system of the Lund University and the results are shown in the analysis bulletin issued on the date 14 December 2001.

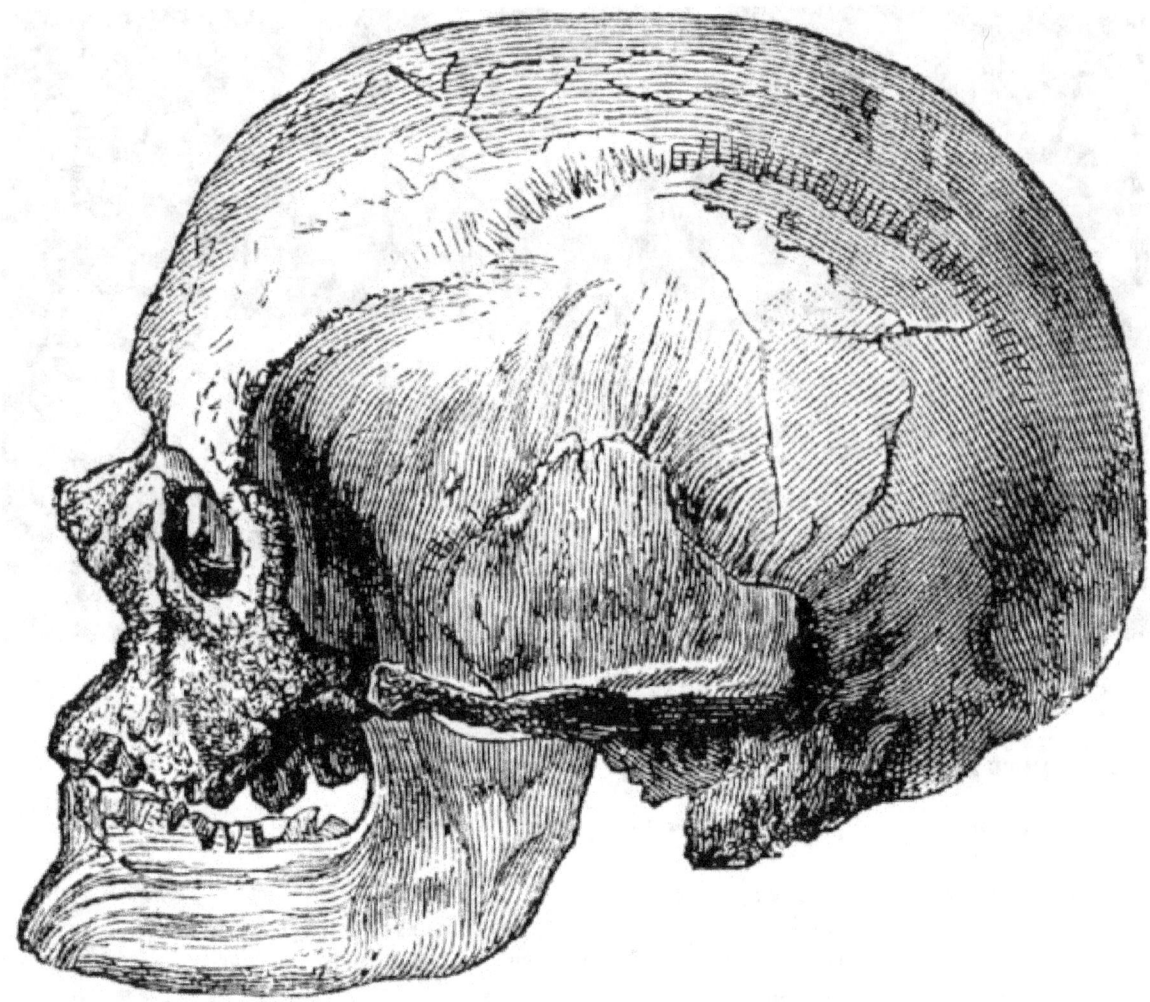

Male Cro-Magnon skull

The human fossil remains from Muierii Cave, Baia de Fier, have been dated to 30,150 ± 800 years BP, and the skull from the Cioclovina Cave has been dated to 29,000 ± 700 years BP.[*][10][*][11][*][12]

14.2   Mesolithic

(9,500 – 7,500 BP)

See also: Mesolithic Europe and Mesolithic Transylvania
 The Mesolithic period began at the end of the Pleistocene epoch (10th millennium BC) and ended with the Neolithic introduction of farming, the date of which varied in each geographical region. According to Douglass W. Bailey:[*][13]

The Mesolithic is the transitional period between the Upper Palaeolithic hunter-gathering existence and the development of farming and pottery production during the Postglacial Neolithic. The duration of the classical Palaeolithic, which lasted until about 10,000 years ago, is applicable to the Balkans. It ended with the Mesolithic (duration is two to four millennia) or, where an early Neolithisation was peculiar to, with the Epipalaeolithic.

Regions with limited glacial impact (e.g. the Balkans), the term Epipalaeolithic is more preferable. Regions that experienced less environmental effects during the last ice age have a much less apparent, straightforward, and occasionally marked by an absence of sites from the Mesolithic era. See the above Douglass W. Bailey quote.

Sculpture found at the archaeological site of Lepenski Vir

There is lithic evidence in Serbia (see Lepenski Vir), southwestern Romania, and Montenegro. At Ostrovul Banului, the Cuina Turcului rock shelter in the Danube Gorges and in the nearby caves of Climente people make relatively advanced bone and lithic tools (i.e. end-scrapers, blade lets, and flakes).

The single site representing materials related to Mesolithic in Bulgaria is Pobiti Kamini. There is no another lithic evidence on the period. There is a 4,000-gap between the latest Upper Palaeolithic material (13,600 BP at Temnta Dupka) and the earliest Neolithic evidence presented at Gulubnik (the beginning of the 7th millennium BC).

At Odmut in Montenegro there is evidence for human activity in the period. The research of the period was supplemented with Greek Mesolithic well represented by sites such as Frachthi Cave. The other sites are Theopetra Cave and Sesklo in Thessaly that represent the Middle and Upper Palaeolithic as well as the early Neolithic period. Yet southern and coastal sites Greece, which contained materials from the Mesolithic are less known.

Activities began to be concentrated around individual sites where people displayed personal and group identities using various decorations: wearing ornaments and painting their bodies with ochre and hematite. As regards the point of identity D. Bailey writes, "Flint-cutting tools as well as time and effort needed to produce such tools testify the expressions of identity and more flexible combinations of materials, which began to be used in the late Upper Palaeolithic and Mesolithic."

The aforementioned allows us to speculate whether or not there was a period which could be described as Mesolithic in southeastern Europe, rather than an extended Upper Palaeolithic. On the other hand, lack of research in a number of regions, and the fact that many of the sites were close to the shore (it is evident that the current sea level is 100 m higher, and a number of sites were covered by water) means that Mesolithic Balkans could be referred to as Epipalaeolithic) Balkans which would better describe its gradual continuity and poorly defined development.

The relative climatic stability in the Balkans, compared to northern and western Europe, enabled continuous settlement in the Balkans, thus effectively functioning as an ice-age refuge from where much of Europe, especially eastern Europe, was re-populated.

14.3 Neolithic

See also: Neolithic Europe, Neolithic Transylvania and Chalcolithic Europe
The Balkans were the site of major Neolithic cultures, including Vinča, Varna, Karanovo, Hamangia.

The Vinča culture was an early culture of the Balkans (between the 6th and the 3rd millennium BC), stretching around the course of the Danube in Serbia, Croatia, Romania, Bulgaria, Montenegro, Kosovo, Albania the Republic of Macedonia, although traces of it can be found all around the Balkans, parts of Central Europe and Asia Minor.

"Kurganization" of the eastern Balkans (and the Cucuteni-Trypillian culture adjacent to the north) during the Eneolithic is associated with an early expansion of Indo-Europeans.

- Butmir culture

- Starčevo-Criş culture

- Dudeşti culture

- Cucuteni-Trypillian culture

- Hamangia culture

- Vinča culture

- Varna culture

- Tărtăria tablets

- Kurgan hypothesis

14.4 Bronze Age

(3,500 – 1,100 BC)

See also: Bronze Age Europe, Bronze Age Romania and Bronze Age Transylvania
The Bronze Age in the Balkans is divided as follows (Boardman p. 166)

- Early Bronze Age: 20th to 16th centuries BC

- Middle Bronze Age: 16th to 14th centuries BC

- Late Bronze Age: 14th to 13th centuries BC

The Bronze Age in the Central and Eastern Balkans begins late, around 1800 BC. The transition to the Iron Age gradually sets in over the 13th century BC.

The "East Balkan Complex" (Karanovo VII, Ezero culture) covers all of Thrace. The Bronze Age cultures of the Central and Western Balkans are less clearly delineated and stretch to Pannonia, the Carpathians and into Hungary.

14.5 Iron Age

(1,100 BC – 150 AD)

See also: Iron Age Europe, Iron Age Romania, Iron Age Transylvania, Thracians, Dacians, Illyrians and Thraco-Cimmerian
After the period that followed the arrival of the Dorians, known as the Greek Dark Ages or Submycenaean Period, the classical Greek culture began to develop in the southern Balkan peninsula, the Aegean islands and the western Asia Minor Greek colonies starting around the 9–8th century (the Geometric Period) and peaking with the 5th century BC Athens democracy.

The Greeks were the first to establish a system of trade routes in the Balkans and, in order to facilitate trade with the natives between 700 BC and 300 BC, they founded several colonies on the Black Sea (Pontus Euxinus) coast, Asia Minor, Dalmatia, Southern Italy (Magna Graecia) etc.

The other peoples of the Balkans organized themselves in large tribal unions such as the Thracian Odrysian kingdom in the Eastern Balkans in the 5th century BC, and the Illyrian kingdom in the Western Balkans from the early 4th century.

Other tribal unions existed in Dacia at least as early as the beginning of the 2nd century BC under King Oroles. The Illyrian tribes were situated in the area corresponding to today's former Yugoslavia and Albania. The name *Illyrii* was originally used to refer to a people occupying an area centred on Lake Skadar, situated between Albania and Montenegro (see List of ancient tribes in Illyria). The term *Illyria* was subsequently used by the Greeks and Romans as a generic name to refer to different peoples within a well defined but much greater area.[14]

Hellenistic culture spread throughout the Macedonian Empire created by Alexander the Great from the later 4th century BC. By the end of the 4th century BC Greek language and culture were dominant not only in the Balkans but also around the whole Eastern Mediterranean.

By the 6th century BC the first written sources dealing with the territory north of the Danube appear in Greek sources. By this time the Getae (and later the Daci) had branched out from the Thracian-speaking populations.

14.6 See also

- Aegean civilization

- Bronze Age Europe

- Dacia

- History of Eurasia

- History of Europe

- Illyria

- Iron Age Europe

- Mesolithic Europe

- Neolithic Europe

- Old European culture

- Paleo-Balkans languages

- Paleolithic Europe

- Prehistory of Transylvania

- Bronze Age in Romania

- Prehistoric Croatia

- Prehistoric Europe

- Prehistoric Serbia

- Prehistory

- Proto-Indo-Europeans

- Stone Age

- Thracia

- Thracian language

- Timeline of glaciation

14.7   References

Inline

[1] Classical World

[2] e.g. Thrace in book V.

[3] http://www.academia.edu/400095/Sirakov_et_al._2010_.-_An_ancient_continuous_human_presence_in_the_Balkans_and_the_
     beginnings_of_human_settlement_in_western_Eurasia_A_Lower_Pleistocene_example_of_the_Lower_Palaeolithic_levels_in_
     Kozarnika_cave_North-western_Bulgaria_

[4] Balkan prehistory Page 15 By Douglass W. Bailey ISBN 0-415-21597-8

[5] Trinkaus, E., Milota, Ş., Rodrigo, R., Gherase, M., Moldovan, O. (2003), Early Modern Human Cranial remains from the
     Peştera cu Oase, Romania in *Journal of Human Evolution*, **45**, pp. 245 –253,

[6] João Zilhão, (2006), Neanderthals and Moderns Mixed and It Matters, in *Evolutionary Anthropology*, **15**:183–195, p.185

[7] Trinkaus, E., Moldovan, O., Milota, Ş., Bîlgăr, A., Sarcina, L., Athreya, S., Bailey, S.E., Rodrigo, R., Gherase, M., Hilgham, T., Bronk Ramsey, C., & Van Der Plicht, J. (2003), An early modern human from Peştera cu Oase, Romania. *Proceedings of the National Acadademy of Science U.S.A.*, **100**(20), pp. 11231–11236

[8] Andrei Soficaru, Adrian Dobo and Erik Trinkaus (2006), Early modern humans from the Peştera Muierii, Baia de Fier, Romania, *Proceedings of the National Acadademy of Science U.S.A.*, **103**(46), pp. 17196-17201

[9] Harvati K, Gunz P, Grigorescu D. Cioclovina (Romania): affinities of an early modern European. J Hum Evol. 2007 Dec;53(6):732-46

[10] Olariu A., Alexandrescu E., Skog G., Hellborg R., Stenström K., Faarinen M. and Persson P, Dating of two Palaeolithic human fossil bones from Romania by accelerator mass spectrometry, NIPNE Scientific Reports 2001-202, pag. 82

[11] Olariu A., Skog G., Hellborg R., Stenström K., Faarinen M. and Persson P. and Alexandrescu E., 2003, Dating of two Palaeolithic human fossil bones from Romania by accelerator mass spectrometry, http://arXiv.org/abs/physics/0309110

[12] Olariu A., Stenström K. and Hellborg R. (Eds), 2005, Proceedings of International conference on Applications of High Precision Atomic & Nuclear Methods, 2–6 September 2002, Neptun, Romania, Publishing House of Romanian Academy, Bucharest, ISBN 973-27-1181-7, Dating of two Palaeolithic human fossil bones from Romania by accelerator mass spectrometry, 235-239

[13] Balkan prehistory Page 36 By Douglass W. Bailey ISBN 0-415-21597-8

[14] The Illyrians. John Wilkes

General

- John Boardman, *The Cambridge Ancient History*, Part I: The Prehistory of the Balkans to 1000 BC, Cambridge University Press (1923), ISBN 0-521-22496-9.

- Douglass W. Bailey, *Balkan prehistory*, ISBN 0-415-21597-8

- Alexandru Păunescu, *Evoluţia istorică pe teritoriul României din paleolitic până la inceputul Neoliticului*, SCIVA, 31, 1980, 4, p. 519-545.

- Paul Lachlan MacKendrick, *The Dacian Stones Speak*, The University of North Carolina Press, Chapel Hill, 1974, ISBN 0-8078-4939-1

14.8 External links

- Periodization of Balkan Prehistory ~ 6200 - 1100 BC

- South East Europe pre-history summary to 700BC

- Balkan Prehistory: Exclusion, Incorporation and Identity by Douglass W. Bailey

- The Aegeo-Balkan Prehistory Project

- (Romanian) Ion din Anina, primul om din Europa on Jurnalul.ro

- (English) Human fossils set European record on BBC.co.uk

- (Romanian) Enciclopedia României

The Thinker of Hamangia, *Neolithic Hamangia culture (c. 5250-4550 BC)*

The Bronze Age hoard of Cugir, 12th century BC, in display at the National Museum of the Union, Alba Iulia

Wietenberg culture battle axes found at Valea Chioarului, Maramureş County, Romania. In display at the National Museum of Transylvanian History, Cluj-Napoca

The Helmet of Coţofeneşti - a full gold Geto-Dacian helmet dating from the first half of the 4th century BC, currently at the National Museum of Romanian History

Distribution of "Thraco-Cimmerian" finds

Chapter 15

Bronze Age in Romania

See also: Bronze Age in Transylvania, Bronze Age in Southeastern Europe and Bronze Age in Europe
 The Bronze Age is a period in the Prehistoric Romanian timeline and is sub-divided into Early Bronze Age (ca. 3500–

Late Bronze Age vessels and tools, from various locations: Alba Iulia, Straja, Geoagiu de Sus, Piatra Craivii. In display at the National Museum of the Union, Alba Iulia

2200 BCE), Middle Bronze Age (ca.2200–1600/1500 BCE), and Late Bronze Age (ca. 1600/1500–1100 BCE).[*][1]

15.1   Periodization

Several Bronze Age chronologies have been applied to the Romanian area. An example would be the Periodization of Paul Reinecke for the Central European space, which split the Bronze Age into four phases (A, B, C and D) based upon the associations among the found bronze objects.*[1]

15.2   Features

Coțofeni culture pottery at the Aiud History Museum, AiudThe Neolithic society is characterized by a predominantly agrarian economy with stable settlements and funerary practices specific to a religion based on a fertility cult. This type of society is near the peer polity described by Colin Renfrew.[1] One of the most important Neolithic archaeological cultures in the Romanian territory is the Cucuteni culture, one of the oldest in Europe (see Old Europe). This culture is probably the last that created painted pottery in Europe.*

During the Bronze Age, there were some important developments from Chalcolithic, with significant improvements in the economy.

The local bronze-aged economy was based on rearing livestock (sheep, goats and pigs). The Wietenberg culture reared large cattle and horses for both transportation and food. At this time, the artistic output also significantly increased, for example the Gârla Mare culture who created intricate clay statuettes.

In the Early Bronze Age (ca. 3500–2200 BCE), we see the archaeological evidence of various cultures developing, including the Baden-Coțofeni culture, the Cernavodă III-Belleraz culture, the Glina culture and the Verbicioara culture. Common occupations were agriculture, mining, and animal husbandry. Houses were rectangular and medium-sized. The last period of the Early Bronze Age produced a broad range of ornaments (loop rings, bracelets, necklaces, pendants comprising copper, gold, and silver and particularly bronze).

Verbicioara culture was identified in 1949 by the eponymous resort excavations. Regarding burial customs, it was considered the beginning of the burial of the dead.[*][2]

In the Middle Bronze Age (ca.2200–1600/1500 BCE), the population of Romania and neighboring countries was demarcated by the appearance of several major cultures. Some that stand out include the Otomani culture (seen also in Slovakia), Wietenberg culture (seen in Transylvania), Mureş culture, and Gârla Mare culture (from which impressive clay figurines and statuettes have been found).

15.3 Religion

Wietenberg culture artifacts. **In display at the National Museum of Transylvanian History, Cluj-Napoca**

The Bronze Age introduced solar, or Uranian, cults. Some ornaments, considered to be solar symbols, were frequently pictured on ceramic or metal parts: concentric circles, circles accompanied by rays, and the swastika. Cremation is considered to be connected to these cults.[*][3]

In the Romanian territory, there are three known bronze-aged sanctuaries: Sălacea, Bihor County (Ottomány culture, phase II), . The only cultures of this area well represented in this regard are the Gârla Mare Zuto Brdo culture and the Bijelo Szeremle Brdo-Dalj culture (also present in Hungary and Croatia). About 340 pieces were found in the area of the two cultures, of which 244 are in the Gârla Mare area.[*][3]

Clay miniature axes (axes, hammers or double axes) belonging to this period have been found. Labrys double-axes are frequently found in the Cretan and Mycenaean worlds, where they occur most often in complex rituals and tombs (for example the *Tomb of double ax of Knossos*). In the Mycenaean context, the labrys has a wide range of sizes, from miniature forms to giant forms that measure 1.20 meters. However, the labrys site is frequently associated with the moon and can be a symbol of a goddess of vegetation, the forerunner of Demeter, who, on Mycenaean seals, is found under a tree. The goddess has an ax in her hand and receives as gifts poppies and fruits.

15.4 See also

- Bronze Age in Transylvania

- Bronze Age in Southeastern Europe

- Bronze Age in Europe

- Basarabi culture

- Coțofeni culture

- Otomani culture

- Pecica culture

- Wietenberg culture

- Celts in Transylvania

- Getae

- Rotbav Archaeological Site

15.5 Notes

[1] Cristian Ştefan-*Epoca Bronzului*, page 1

[2] "Cu Privire La Descoperirile Funerare Ale Grupei Verbicioara". Archaeology.ro. 2004-12-09. Retrieved 2012-02-09.

[3] "Credinte religioase si piese de cult in epoca bronzului - Prehistoire". Prehistoire.e-monsite.com. Retrieved 2012-02-09.

15.6 References

- Cristian Ştefan - *Epoca Bronzului*

- Ioan Aurel-Pop, Ioan Bolovan, coordinatings - *Istoria ilustrată a României*

15.7 External links

- http://prehistoire.e-monsite.com/rubrique,epoca-bronzului-ii,1112106.html

- http://www.archaeology.ro/imc_verb.htm

- http://www.archaeology.ro/imc_mont.htm

Chapter 16

Atlantic Bronze Age

The **Atlantic Bronze Age** is a cultural complex of the Bronze Age period of approximately 1300–700 BC that includes different cultures in Portugal, Andalusia, Galicia, Armorica and the British Isles.

16.1 Trade

The Atlantic Bronze Age is marked by economic and cultural exchange that led to the high degree of cultural similarity exhibited by the coastal communities from Central Portugal to Galicia, Armorica and Scotland, including the frequent use of stones as chevaux-de-frise, the establishment of cliff castles, or the domestic architecture sometimes characterized by the round houses.*[1] Commercial contacts extended from Sweden*[2] and Denmark to the Mediterranean.*[1] The period was defined by a number of distinct regional centres of metal production, unified by a regular maritime exchange of some of their products. The major centres were southern England and Ireland, north-western France, and western Iberia.*[3]

The items related to this culture are frequently found forming hoards, or they are deposited in ritual areas,*[4]*[5] usually watery contexts: rivers, lakes and bogs. Among the more noted items belonging to this cultural complex we can count the socketed and double ring bronze axes, sometimes buried forming large hoards in Brittany and Galicia; war gear, as lunate spearheads, V-notched shields, and a variety of bronze swords —among them carp's-tongue ones—usually found deposited in lakes, rivers or rocky outcrops;*[6] and the elites feasting gear: articulated roasting spits, cauldrons, and flesh hooks,*[5]*[7] found from central Portugal to Scotland.*[1]

The origins of the Celts were attributed to this period in 2008 by John T. Koch*[8] and supported by Barry Cunliffe,*[9] who argued for the past development of Celtic as an Atlantic lingua franca, later spreading into mainland Europe.*[5] They argue that communities adopted early Late Bronze Age Urnfield (Bronze D and Hallstatt A) elite status markers such as grip-tongue swords and sheet-bronze metalwork, along with new specialist know-how needed for their production and ritual knowledge about their 'proper' treatment upon deposition.*[10] which they see as indicating possible processes linked to language shift.*[10] In 2013, Koch saw this east to west elite contact as the simplest explanation for the genesis of Celtic languages with a Proto-Celtic homeland in west-central Europe.*[11] However, this stands in contrast to what remains the more generally accepted view that Celtic origins lie with the Central European Hallstatt C culture.

16.2 See also

- Atlantic Europe
- Bronze Age Britain
- Bronze Age Europe
- Castro Culture

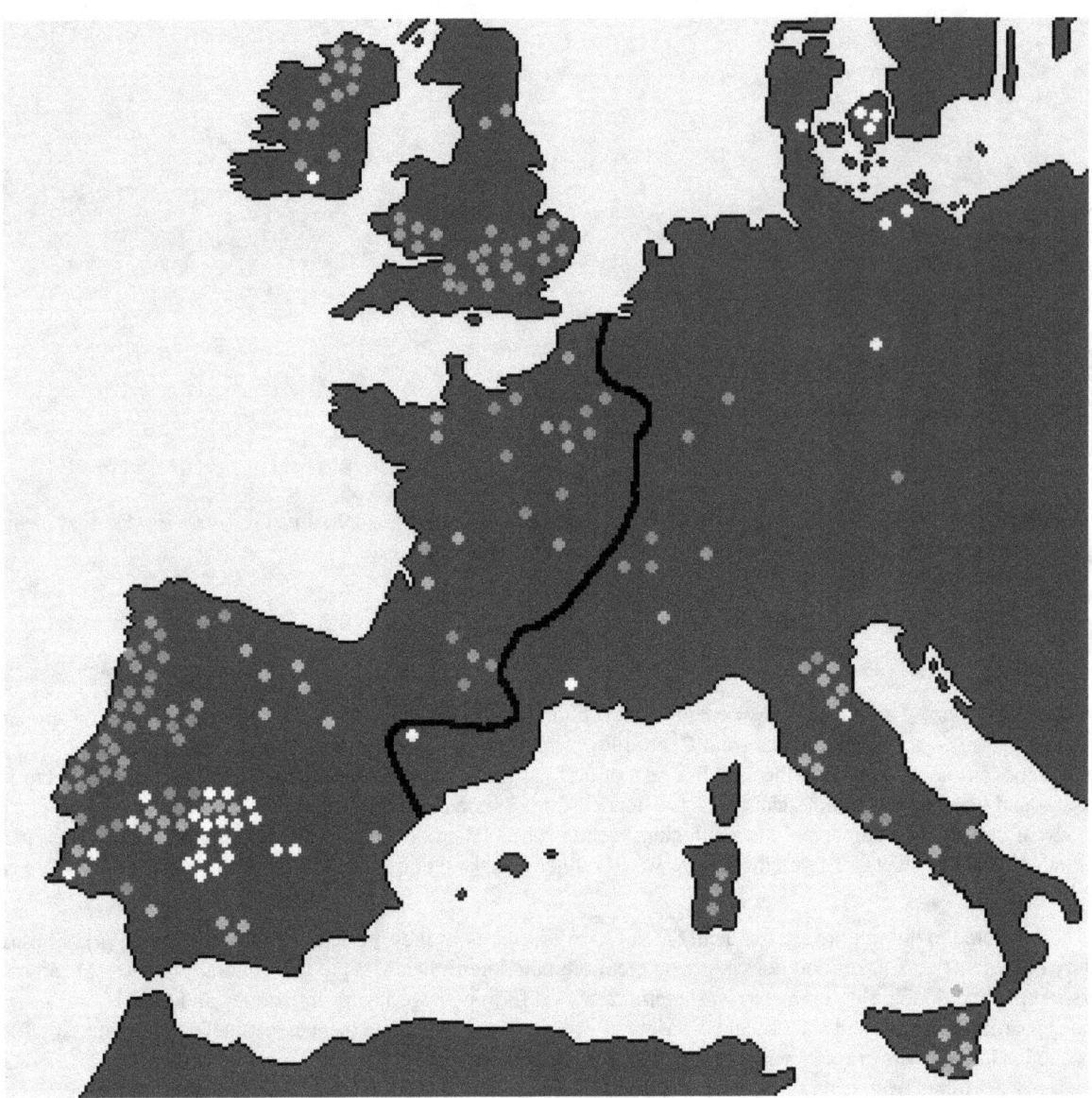

Atlantic Bronze Age exchange products:

Imports:
- Urnfields' crested helmets
- "Elbow" fibulae
  "Cleavage" shields

Exports:
- Tube sickles (Britain)
- Tube and double ring axes (Portugal)

▬▬▬ Approx. western limit of Urnfields expansion

- Cogotas
- Early history of Ireland
- Haplogroup R1b (Y-DNA) subclade R1b1b2 (R-M269)
- Haplogroup I (Y-DNA) subclade I-M26

- Haplogroup E (Y-DNA) subclade E1b1b1a (formerly E3b1a2)

- Megaliths

- Prehistoric Iberia

- Vasconic substratum theory

16.3 References

[1] Cunliffe, Barry (1999). "Atlantic Sea-ways" (PDF). *Revista de Guimarães*. Especial (I): 93–105.

[2] Ling, Johan; Stos-Gale, Zofia; Grandin, Lena; Billström, Kjell; Hjärthner-Holdar, Eva; Persson, Per-Olof. "Moving metals II: provenancing Scandinavian Bronze Age artefacts by lead isotope and elemental analyses". *Journal of Archaeological Science* **41**: 106–132. doi:10.1016/j.jas.2013.07.018.

[3] Europe Before History by Kristian Kristiansen

[4] Comendador Rey, Beatriz. "SPACE AND MEMORY AT THE MOUTH OF THE RIVER ULLA (GALICIA, SPAIN)" (PDF). *Conceptualising Space and Place: On the role of agency, memory and identity in the construction of space from the Upper Palaeolithic to the Iron Age in Europe*. Archaeopress. Retrieved 26 April 2011.

[5] Cunliffe, Barry (2008). *Europe between the oceans : themes and variations, 9000 BC-AD 1000* (First printed in paperback 2011. ed.). New Haven: Yale University Press. pp. 254–258. ISBN 978-0-300-17086-3.

[6] Quilliec, Bénédicte T. (2007). "Life and death of an Atlantic sword: Reconstruction of the processes of fabrication, use wear and destruction" (PDF). *Complutum* **18**: 93–107. Retrieved 22 September 2011.

[7] Bowman, Sheridan; Stuart Needham (2007). "THE DUNAVERNEY AND LITTLE THETFORD FLESH-HOOKS: HISTORY, TECHNOLOGY AND THEIR POSITION WITHIN THE LATER BRONZE AGE ATLANTIC ZONE FEASTING COMPLEX" (PDF). *The Antiquaries Journal* **87**: 53–108. doi:10.1017/s0003581500000846. Retrieved 22 September 2011.

[8] Koch, John (2009). *Tartessian: Celtic from the Southwest at the Dawn of History in Acta Palaeohispanica X Palaeohispanica 9 (2009)* (PDF). Palaeohispanica. pp. 339–351. ISSN 1578-5386. Archived (PDF) from the original on 23 June 2010. Retrieved 2010-05-17.

[9] Cunliffe, Barry (2008). *A Race Apart: Insularity and Connectivity in Proceedings of the Prehistoric Society 75, 2009, pp. 55–64*. The Prehistoric Society. p. 61.

[10] Brandherm, Dirk (2013). *Celtic from the West 2 - Westward Ho? Sword-bearers and all the rest of it...* Oxford: Oxbow Books. p. 148. ISBN 978-1-84217-529-3.

[11] Koch, John T. (2013). *Celtic from the West 2 -Prologue: The Earliest Hallstatt Iron Age cannot equal Proto-Celtic*. Oxford: Oxbow Books. p. 10. ISBN 978-1-84217-529-3.

16.4 External links

- Spaniards search for legendary Tartessos in a marsh

- Divers unearth Bronze Age hoard off the coast of Devon

- Moor Sands finds, including a remarkably well preserved and complete sword which has parallels with material from the Seine basin of northern France

- 3000 year old shipwreck shows European trade was thriving in Bronze Age

Chapter 17

Bronze Age Britain

Bronze Age Britain refers to the period of British history that spanned from c. 2500 until c. 800 BC.[*][1] Lasting for approximately 1,700 years, it was preceded by the era of Neolithic Britain and was in turn followed by the era of Iron Age Britain. Being categorised as a Bronze Age, it was marked by the use of copper and then bronze by the prehistoric Britons, who used such metals to fashion tools. Great Britain in the Bronze Age also saw the widespread adoption of agriculture.

During the British Bronze Age, large megalithic monuments similar to those from the Late Neolithic continued to be constructed or modified, including such sites as Avebury, Stonehenge and Silbury Hill. This has been described as a time "when elaborate ceremonial practices emerged among some communities of subsistence agriculturalists of western Europe".[*][2]

17.1 History

17.1.1 Early Bronze Age (EBA), c. 2500-1500 BC

There is no clear consensus on the date for the beginning of the Bronze Age in Great Britain and Ireland. Some sources give a date as late as 2000 BC,[*][3] while others set 2200 BC as the demarcation between the Neolithic and the Bronze Age.[*][4] The period from 2500 BC to 2000 BC has been called the "Late Neolithic/Early Bronze Age", in recognition of the difficulty of exactly defining this boundary.[*][5]

- 2500-2000 BC: Mount Pleasant Phase, Early Beaker culture: ; Britain: copper+tin.

- 2100-1900 BC: Late Beaker: knives, tanged spearheads (Bush Barrow; Overton Period).

- 1800-1600 BC: Fargo Phase (see correction at Bedd Branwen Period); burials.

17.1.2 Middle Bronze Age (MBA), 1500-1000 BC

- 1500-1300 BC: Acton Park Phase: palstaves, socketed spearheads; copper+tin, also lead.

- 1300-1200 BC: Knighton Heath Period; "rapiers."

- 1200-1000 BC: Early Urnfield; Wilburton-Wallington Phase.

17.1.3 Late Bronze Age (LBA), 1000-700 BC

- 1000-900 BC: Late Urnfield: socketed axes, palstaves (also lead).

Bronze shield, 1200-700 BC

- 800-700 BC: Ewart Park Phase, Llyn Fawr Phase: leaf-shaped swords.

17.2 Development

17.2.1 The Beaker culture

In around 2700 BC a new pottery style arrived in Great Britain, often referred to as the Beaker culture. Beaker pottery appears in the Mount Pleasant Phase (2700 - 2000 BC), along with flat axes and burial practices of inhumation. People of this period were also largely responsible for building many famous prehistoric sites, such as the later phases of Stonehenge along with Seahenge.

Movement of Europeans brought new people to the islands from the continent. Recent tooth enamel isotope research on bodies found in early Bronze Age graves around Stonehenge indicates that at least some of the new arrivals came from

Socketed axes from a hoard

the area of modern Switzerland. The Beaker culture displayed different behaviours from the earlier Neolithic people and cultural change was significant. Integration is thought to have been peaceful, as many of the early henge sites were seemingly adopted by the newcomers.

Also, the burial of dead (which until this period had usually been communal) became more individual. For example, whereas in the Neolithic a large chambered cairn or long barrow was used to house the dead, the 'Early Bronze Age' saw people buried in individual barrows (also commonly known and marked on modern British Ordnance Survey maps as tumuli) or sometimes in cists covered with cairns. They were often buried with a beaker alongside the body.

There is some debate amongst archaeologists as to whether the 'Beaker people' were a race of people who migrated to Great Britain and Ireland *en masse* from the continent, or whether a prestigious Beaker cultural "package" of goods and behaviours (which eventually spread across most of western Europe) diffused to the islands' existing inhabitants through trade across tribal boundaries. Modern thinking tends towards the latter view. Alternatively, a Beaker elite may have made the migration and come to influence the native population at some level.

17.2.2   Bronze

Believed to be of Iberian origin (modern day Spain and Portugal), part of the Beaker culture brought to Great Britain the skill of refining metal. At first they made items from copper, but from around 2150 BC smiths had discovered how to make bronze (which is much harder than copper) by mixing copper with a small amount of tin. With this discovery, the Bronze Age began in Great Britain. Over the next thousand years, bronze gradually replaced stone as the main material for tool and weapon making.

The bronze axehead, made by casting, was at first similar to its stone predecessors but then developed a socket for the wooden handle to fit into, and a small loop or ring to make lashing the two together easier. Groups of unused axes are often found together, suggesting ritual deposits to some, though many archaeologists believe that elite groups collected bronze items, perhaps restricting their use among the wider population. Bronze swords of a graceful "leaf" shape, swelling gently from the handle before coming to a tip, have been found in considerable numbers, along with spear heads and arrow points.

Great Britain had large reserves of tin in the areas of Cornwall and Devon in what is now Southwest England, and thus tin mining began. By around 1600 BC, the southwest of the island was experiencing a trade boom as British tin was exported across Europe.

Bronze-age Britons were also skilled at making jewellery from gold, as well as occasional objects like the Rillaton Cup and Mold Cape. Many examples of these have been found in graves of the wealthy Wessex culture of Southern Britain, though they are not as frequent as Irish finds.

The greatest quantities of bronze objects found in what is now England were discovered in East Cambridgeshire, where the most important finds were recovered in Isleham (more than 6500 pieces).*[6]

The earliest known metalworking building was found at Sigwells, Somerset, England. Several casting mould fragments were fitted to a Wilburton type sword held in Somerset County Museum.*[7] They were found in association with cereal grain dated to the 12th century BC by carbon dating.

17.2.3   The Wessex culture

The rich Wessex culture developed in southern Great Britain at this time. The weather, previously warm and dry, became much wetter as the Bronze Age continued, forcing the population away from easily defended sites in the hills and into the fertile valleys. Large livestock farms developed in the lowlands which appear to have contributed to economic growth and inspired increasing forest clearances.

17.2.4   The Deverel-Rimbury culture

The Deverel-Rimbury culture began to emerge in the second half of the 'Middle Bronze Age' (c. 1400-1100 BC) to exploit the wetter conditions. Cornwall was a major source of tin for much of western Europe and copper was extracted from sites such as the Great Orme mine in Northern Wales. Social groups appear to have been tribal but with growing complexity and hierarchies becoming apparent.

17.2.5   Disruption of cultural patterns

See also: Atlantic Bronze Age

There is evidence of a relatively large-scale disruption of cultural patterns which some scholars think may indicate an invasion (or at least a migration) into Southern Great Britain around the 12th century BC. This disruption was felt far beyond Britain, even beyond Europe, as most of the great Near Eastern empires collapsed (or experienced severe difficulties) and the Sea Peoples harried the entire Mediterranean basin around this time. Cremation was adopted as a burial practice, with cemeteries of urns containing cremated individuals appearing in the archaeological record. According to John T. Koch and others, the Celtic languages developed during this Late Bronze Age period in an intensely trading-networked culture called the Atlantic Bronze Age that included Britain, Ireland, France, Spain and Portugal,*[8]*[9]*[10]*[11]*[12]*[13] but this stands in contrast to the more generally accepted view that Celtic origins lie with the Hallstatt culture.

17.3   Bronze Age seafaring

- Ferriby Boats

- Langdon Bay hoard - see also Dover Museum

- Divers unearth Bronze Age hoard off the coast of Devon

- Moor Sands finds, including a remarkably well preserved and complete sword which has parallels with material from the Seine basin of northern France

- 3000 year old shipwreck shows European trade was thriving in Bronze Age

17.4   See also

- List of Bronze Age hoards in Great Britain

- Copper and Bronze Age Ireland

17.5   References

Footnotes

[1] Adkins, Adkins and Leitch 2008. p. 64.

[2] Barrett 1994. p. 05.

[3] Bradley, *Prehistory of Britain and Ireland*, p. 183.

[4] Pollard, "Construction of Prehistoric Britain", in Pollard (ed.), *Prehistoric Britain*, p. 9.

[5] Prior, *Britain BC*, p. 226.

[6] Hall and Coles, p. 81–88.

[7] Tabor, Richard (2008). *Cadbury Castle: The hillfort and landscapes*. Stroud: The History Press. pp. 61–69. ISBN 978-0-7524-4715-5.

[8] http://www.aber.ac.uk/aberonline/en/archive/2008/05/au7608/

[9] "O'Donnell Lecture 2008 Appendix" (PDF).

[10] Koch, John (2009). *Tartessian: Celtic from the Southwest at the Dawn of History in Acta Palaeohispanica X Palaeohispanica 9 (2009)* (PDF). Palaeohispanica. pp. 339–351. ISSN 1578-5386. Retrieved 2010-05-17.

[11] Koch, John. "New research suggests Welsh Celtic roots lie in Spain and Portugal". Retrieved 10 May 2010.

[12] Cunliffe, Karl, Guerra, McEvoy, Bradley; Oppenheimer, Rrvik, Isaac, Parsons, Koch, Freeman and Wodtko (2010). *Celtic from the West: Alternative Perspectives from Archaeology, Genetics, Language and Literature*. Oxbow Books and Celtic Studies Publications. p. 384. ISBN 978-1-84217-410-4.

[13] "Rethinking the Bronze Age and the Arrival of Indo-European in Atlantic Europe" (PDF). University of Wales Centre for Advanced Welsh and Celtic Studies and Institute of Archaeology, University of Oxford. Retrieved 24 May 2010.

Bibliography

- Barrett, John C. (1994). *Fragments from Antiquity: An Archaeology of Social Life in Britain, 2900-1200 BC*. Oxford and Cambridge, Massachusetts: Blackwell.

- Bradley, Richard (2007). *The Prehistory of Britain and Ireland*. Cambridge: Cambridge University Press. ISBN 978-0-521-61270-8.

- Adkins, Roy; Adkins, Lesley; Leitch, Victoria (2008). *The Handbook of British Archaeology (Second Edition).* London: Constable.

- Pearson, Michael Parker (2005). *Bronze Age Britain* (Revised Edition). London: B.T. Batsford. ISBN 0-7134-8849-2.

- Pollard, Joshua (ed.) (2008). *Prehistoric Britain.* Oxford: Blackwell Publishing. ISBN 978-1-4051-2546-8.

- Pryor, Francis (2003). *Britain BC.* London: Harper. ISBN 978-0-00-712693-4.

- R.F. Tylecote, *The early history of metallurgy in Europe* (1987)

17.6 External links

- From Rapier to Langsax: Sword Structure in the British Isles in the Bronze and Iron Ages by Niko Silvester (1995)

Swords found in Scotland

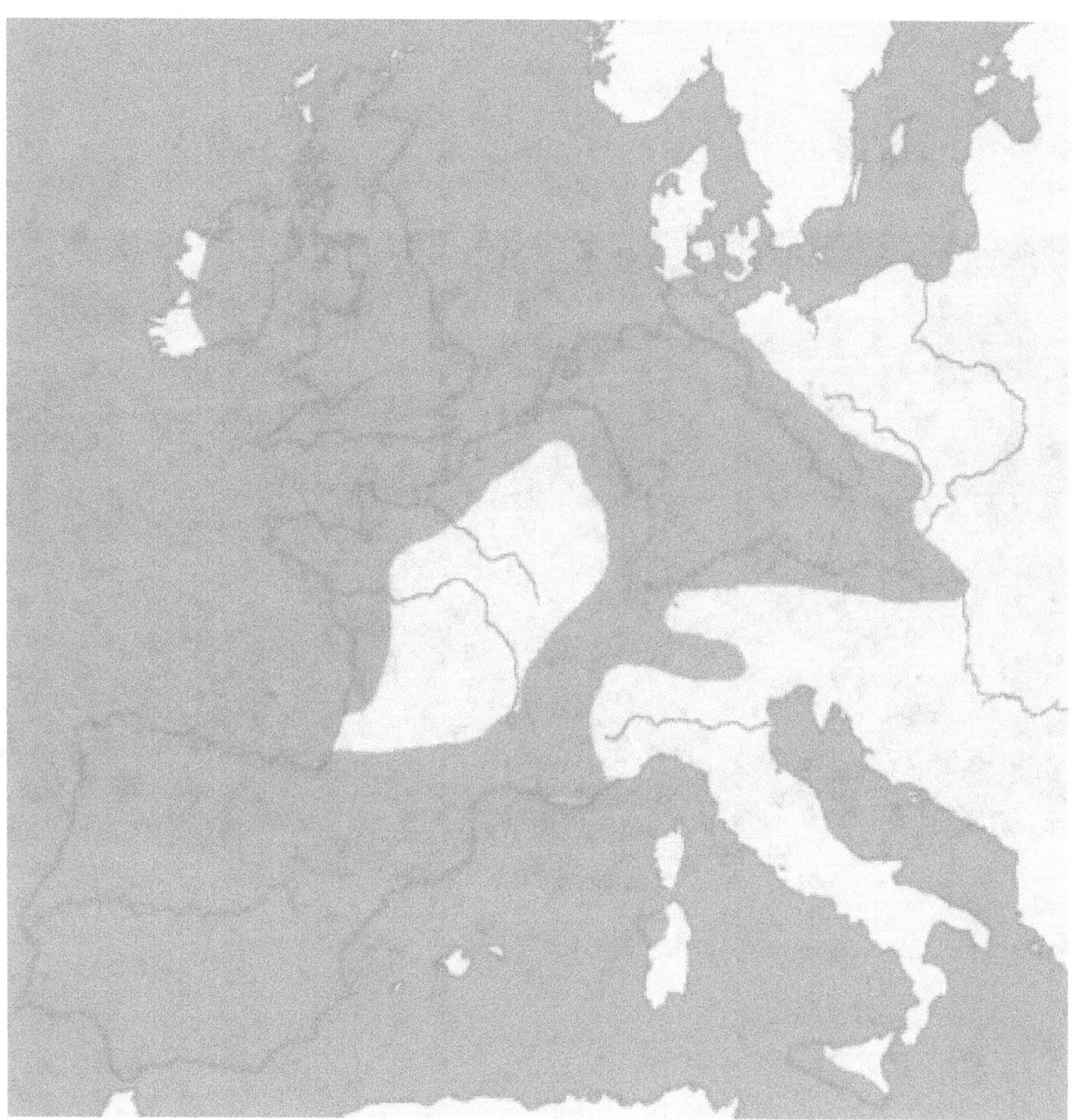

Extent of the Beaker culture

The Mold Cape is unique among survivals

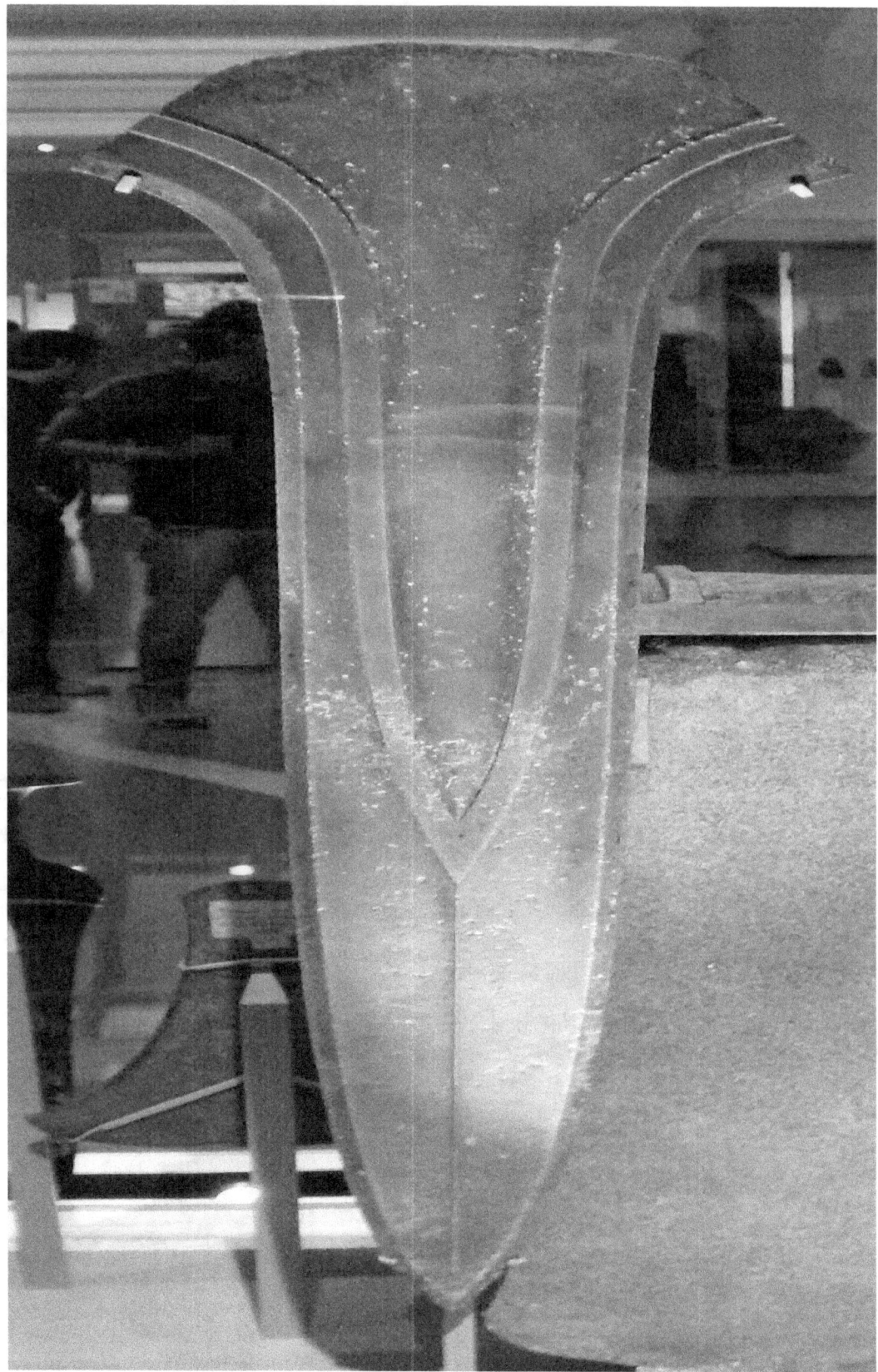

The massive bronze Oxborough Dirk is too large to use

Chapter 18

Nordic Bronze Age

The **Nordic Bronze Age** (also **Northern Bronze Age**) is a period of Scandinavian prehistory from c. 1700–500 BC. The Bronze Age culture of this era succeeded the Late Neolithic Stone Age culture and was followed by the Pre-Roman Iron Age. The archaeological legacy of the Nordic Bronze Age culture is rich, but the ethnic and linguistic affinities of it are unknown, in the absence of written sources. Some scholars also includes sites in what is now northern Germany, Pomerania and Estonia in the Baltic region, as part of its cultural sphere.

18.1 General characteristics

Even though Scandinavians joined the European Bronze Age cultures fairly late through trade, Scandinavian sites presents a rich and well-preserved legacy of bronze and gold objects. These valuable metals were all imported, primarily from Central Europe, but they were often crafted locally and the craftsmanship and metallurgy of the Nordic Bronze Age was of a high standard. The archaeological legacy also comprise locally crafted wool and wooden objects and there are many tumuli and rock carving sites from this period, but no written language existed in the Nordic countries during the Bronze Age. The rock carvings have been dated through comparison with depicted artifacts, for example bronze axes and swords. There are also numerous Nordic Stone Age rock carvings, those of northern Scandinavia, mostly portrays elk.

Thousands of rock carvings from this period depict ships, and the large stone burial monuments known as stone ships, suggest that ships and seafaring played an important role in the culture at large. The depicted ships, most likely represents sewn plank built canoes used for warfare, fishing and trade. These ship types may have their origin as far back as the neolithic period and they continue into the Pre-Roman Iron Age, as exemplified by the Hjortspring boat.[*][2]

18.2 Sub-periodization

Oscar Montelius, who coined the term used for the period, divided it into six distinct sub-periods in his piece *Om tidsbestämning inom bronsåldern med särskilt avseende på Skandinavien* ("On Bronze Age dating with particular focus on Scandinavia") published in 1885, which is still in wide use. His absolute chronology has held up well against radiocarbon dating, with the exception that the period's start is closer to 1700 BC than 1800 BC, as Montelius suggested. For Central Europe a different system developed by Paul Reinecke is commonly used, as each area has its own artifact types and archaeological periods.

A broader subdivision is the Early Bronze Age, between 1700 BC and 1100 BC, and the Late Bronze Age, 1100 BC to 550 BC. These divisions and periods are followed by the Pre-Roman Iron Age.

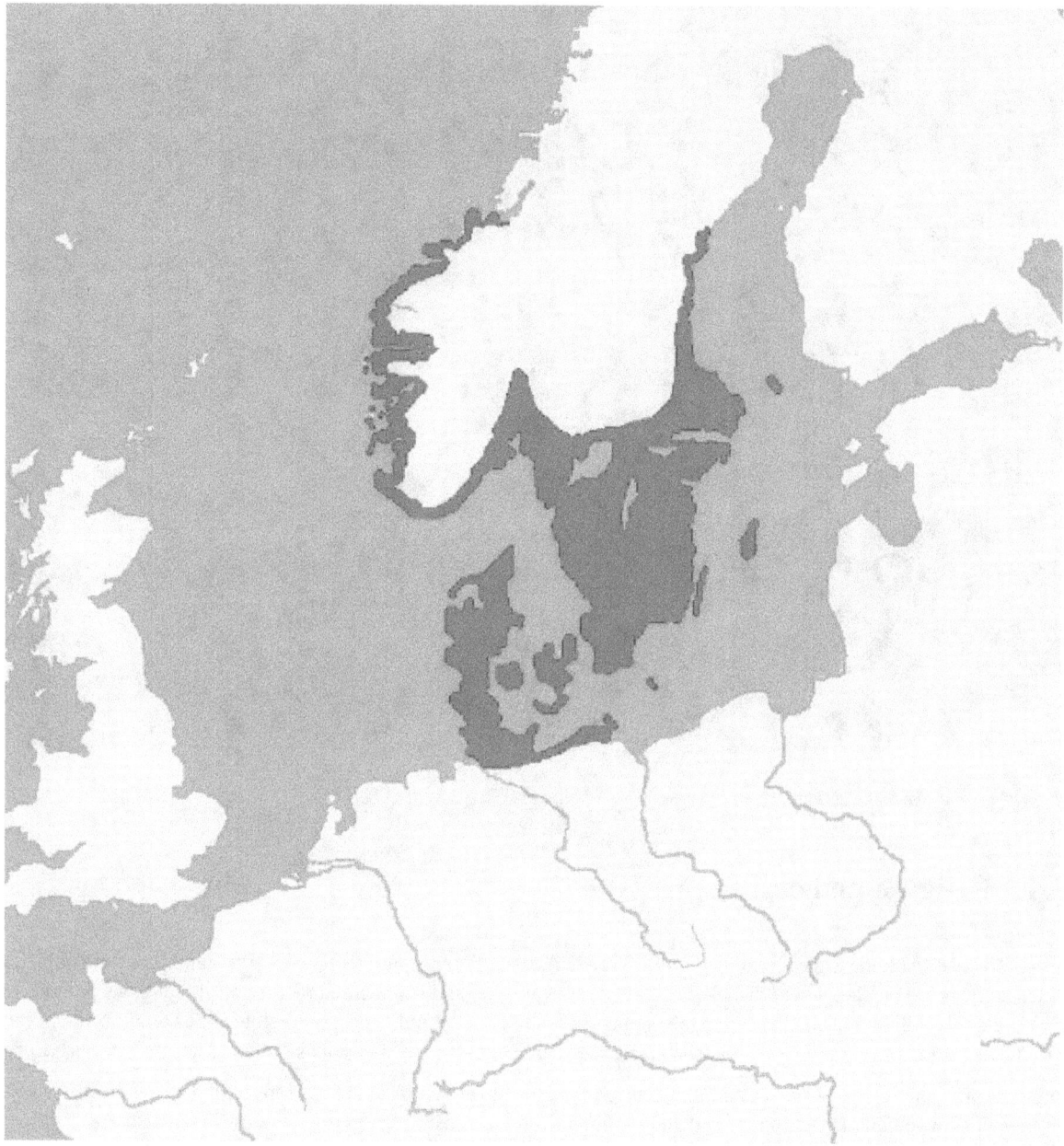

Map of the Nordic Bronze Age culture, c. 1200 BC

18.3 Climate

The Nordic Bronze Age was characterized first by a warm climate that began with a climate change around 2700 BC (comparable to that of present-day central Germany and northern France). The warm climate permitted a relatively dense population and good farming; for example, grapes were grown in Scandinavia at this time. A wetter, colder climate prevailed after a minor change in climate between 850 BC and 760 BC, and a more radical one around 650 BC.

Collage of Bronze Age rock carvings. From Häljesta, Västmanland province, Sweden.[1]

18.4   Religion and cult

There is no coherent knowledge about the Nordic Bronze Age religion; its pantheon, world view and how it was practised. Written sources are lacking, but archaeological finds draws a vague and fragmented picture of the religious practices and the nature of the religion of this period. Only some possible sects and only certain possible tribes are known. Some of the best clues come from tumuli, elaborate artifacts, votive offerings and rock carvings scattered across Northern Europe.

Many finds indicates a strong sun worshipping cult in the Nordic Bronze Age and various animals have been associated with the suns movement across the sky, including horses, birds, snakes and marine creatures (see also *Sól*). A female or mother goddess is also believed to have been widely worshipped (see Nerthus). *Hieros gamos* rites may have been common and there have been several finds of fertility symbols. A pair of twin gods are believed to have been worshipped, and is reflected in a duality in all things sacred: where sacrificial artifacts have been buried they are often found in pairs. Sacrifices (animals, weapons, jewellery and humans) often had a strong connection to bodies of water. Boglands, ponds, streams or lakes were often used as ceremonial and holy places for sacrifices and many artifacts have been found in such locations. Ritual instruments such as bronze lurs have been uncovered, especially in the region of Denmark and western Sweden. Lur horns are also depicted in several rock carvings and are believed to have been used in ceremonies.

Bronze Age rock carvings may contain some of the earliest depictions of well-known gods from the later Norse mythology. A common figure in these rock carvings is that of a male figure carrying what appears to be an axe or hammer. This may have been an early representation of Thor. Other male figures are shown holding a spear. Whether this is a representation of Odin or Týr is not known. It is possible the figure may have been a representation of Tyr, as one example of a Bronze Age rock carving appears to show a figure missing a hand. A figure holding a bow may be an early representation of Ullr. Or it is possible that these figures were not gods at all, but men brandishing the weapons of their culture.

Remnants of the Bronze Age religion and mythology are believed to exist in Germanic mythology and Norse mythology; e.g., Skinfaxi and Hrímfaxi and Nerthus, and it is believed to itself be descended from an older Indo-European prototype.

18.5 Culture

18.6 See also

- Bronze Age Europe

- Bronze Age sword

- Egtved Girl

- The King's Grave

- Stone ships

- Tanumshede

- Pomeranian culture

18.7 Notes

[1] The carvings have been painted in recent times. It is unknown whether they were painted originally. Composite image. Nordic Bronze Age.

[2] Ling 2008. *Elevated Rock Art*. GOTARC Serie B. Gothenburg Archaeological Thesis 49. Department of Archaeology and Ancient History, University of Gothenburg, Goumlteborg, 2008. ISBN 978-91-85245-34-5.

18.8 Bibliography

- Dabrowski, J. (1989) Nordische Kreis un Kulturen Polnischer Gebiete. *Die Bronzezeit im Ostseegebiet. Ein Rapport der Kgl. Schwedischen Akademie der Literatur Geschichte und Alter unt Altertumsforschung über das Julita-Symposium 1986.* Ed Ambrosiani, B. Kungl. Vitterhets Historie och Antikvitets Akademien. Konferenser 22. Stockholm.

- Davidson, H. R. Ellis and Gelling, Peter: *The Chariot of the Sun and other Rites and Symbols of the Northern European Bronze Age.*

- K. Demakopoulou (ed.), *Gods and Heroes of the European Bronze Age*, published on the occasion of the exhibition "Gods and Heroes of the Bronze Age. Europe at the Time of Ulysses", from December 19, 1998, to April 5, 1999, at the National Museum of Denmark, Copenhagen, London (1999), ISBN 0-500-01915-0.

- Demougeot, E. *La formation de l'Europe et les invasions barbares*, Paris: Editions Montaigne, 1969-1974.

- Kaliff, Anders. 2001. *Gothic Connections. Contacts between eastern Scandinavia and the southern Baltic coast 1000 BC – 500 AD.*

- Montelius, Oscar, 1885. *Om tidsbestämning inom bronsåldern med särskilt avseende på Skandinavien.*

- Musset, L. *Les invasions: les vagues germanique*, Paris: Presses universitaires de France, 1965.

Chapter 19

Arsenical bronze

Arsenical bronze is an alloy in which arsenic is added to copper as opposed to, or in addition to tin or other constituent metals, to make bronze. The use of arsenic with copper, either as the secondary constituent or with another component such as tin, results in a stronger final product and better casting behaviour.[1]

Since copper ore is often naturally contaminated with arsenic, the term "arsenical bronze" when used in archaeology is typically only applied to alloys with an arsenic content higher than 1% by weight, in order to distinguish it from potentially accidental additions of arsenic.[2]

19.1 Origins in pre-history

Although arsenical bronze occurs in the archaeological record across the globe, the earliest artifacts so far known have been found on the Iranian plateau in the 5th millennium BCE.[3] Arsenic is present in a number of copper containing ores (see table at right, adapted from Lechtman & Klein, 1999[4]) and therefore some contamination of the copper with arsenic would be unavoidable. However it is still not entirely clear to what extent arsenic was deliberately added to copper [5] and how much its use arose simply from its presence in copper ores that were then treated by smelting to produce the metal.

A possible sequence of events in prehistory involves considering the structure of copper ore deposits, which are mostly sulphides.[6] The surface minerals would contain some native copper and oxidised minerals, but much of the copper and other minerals would have been washed further into the ore body forming a secondary enrichment zone. This includes many minerals such as tennantite, with their arsenic, copper and iron. So the surface deposits would be used first, and with some work deeper sulphidic ores would have been uncovered and worked, and it would have been discovered that the material from this level had better properties.

Using these various ores, there are four possible methods that may have been used to produce arsenical bronze alloys.[3] These are:

- The direct addition of arsenic-bearing metals or ores such as realgar to molten copper.

  This method, although possible, lacks evidence.

- The reduction of antimony-bearing copper arsenates or fahlore to produce an alloy high in arsenic and antimony.

  This is entirely practicable.

- The reduction of roasted copper sulfarsenides such as tennantite and enargite.

  This method would result in the production of toxic fumes of arsenous oxide and the loss of much of the arsenic present in the ores.[7]

- The co-smelting of oxidic and sulphidic ores such as malachite and arsenopyrite together.

  This method has been demonstrated to work well, with little in the way of dangerous fumes given off during it, because of the reactions together among the different minerals.[4]

Furthermore, greater sophistication of metal workers is suggested by Thornton et al.[8] They suggest that iron arsenide was deliberately produced as part of the copper smelting process, to be traded and used to make arsenical bronze elsewhere by addition to molten copper.

Artefacts made of arsenical bronze cover the complete spectrum of metal objects, from axes to ornaments. The method of manufacture involved heating the metal in crucibles and casting it into moulds made of stone or clay. After solidifying it would be polished or in the case of axes and other tools work-hardened by beating the working edge with a hammer, thinning out the metal and increasing its strength.[6] Finished objects could also be engraved or decorated as appropriate.

19.2 Advantages of arsenical bronze

Whilst arsenic was most likely originally mixed with copper as a result of the ores already containing it, its use probably continued for a number of reasons. Firstly, it acts as a de-oxidiser, reacting with oxygen in the hot metal to form arsenous oxides which vaporise from the liquid metal. If a great deal of oxygen is dissolved in liquid copper, when the metal cools the copper oxide separates out at grain boundaries and greatly reduces the ductility of the resulting object. It can lead to a greater risk of porous castings due to the solution of hydrogen in the molten metal and its subsequent loss as a bubble (but any bubbles could be forge welded and still leave the mass of the metal ready to be work-hardened).[1]

Secondly, it is capable of greater work-hardening than is the case with pure copper, so that it performs better when used for cutting or chopping. There is an increase in work-hardening capability with increasing percentage of arsenic, and it can be work-hardened over a wide range of temperatures without fear of embrittlement.[1] Its improved properties over pure copper can be seen with as little as 0.5 to 2 wt% As, giving a 10 to 30% improvement in hardness and tensile strength.[7]

Thirdly, in the correct percentages, it can contribute a silvery sheen to the article being manufactured. There is evidence of arsenical bronze daggers from the Caucasus and other artefacts from different locations having an arsenic rich surface layer which may well have been produced deliberately by ancient craftsmen,[9] and Mexican bells were made of copper with sufficient arsenic to colour them silver.[7]

19.3 Arsenical bronze, sites and civilisations

Arsenical bronze was used by many societies and cultures across the globe. Firstly, the Iranian plateau, followed by the adjacent Mesopotamian area, together covering modern Iran, Iraq and Syria, has the earliest arsenical bronze metallurgy in the world, as previously mentioned. It was in use from the 4th millennium BC through to mid 2nd millennium, a period of nearly 2,000 years. There was a great deal of variation in arsenic content of artefacts throughout this period, making it impossible to say exactly how much was added deliberately and how much came about by accident.[5] Societies using arsenical bronze include the Akkadians, those of Ur, and the Amorites, all based around the Tigris and Euphrates rivers and centres of the trade networks which spread arsenical bronze across the Middle East during the Bronze Age.[5]

The Chalcolithic-period hoard from Nahal Mishmar in the Judean desert west of the Dead Sea contains a number of arsenical bronze (4–12% arsenic) and perhaps arsenical copper artifacts made using the **lost-wax process**, the earliest known use of this complex technique. "Carbon-14 dating of the reed mat in which the objects were wrapped suggests that it dates to at least 3500 B.C. It was in this period that the use of copper became widespread throughout the Levant, attesting to considerable technological developments that parallel major social advances in the region." [10]

Sulfide deposits frequently are a mix of different metal sulfides, such as copper, zinc, silver, lead, arsenic and other metals. (Sphalerite (ZnS_2), for example, is not uncommon in copper sulfide deposits, and the metal smelted would be brass, which is both harder and more durable than copper.)The metals could theoretically be separated out, but the alloys resulting were typically much stronger than the metals individually.

The use of arsenical bronze spread along trade routes into North western China, to the region Gansu – Qinghai, with the Siba, Qijia and Tianshanbeilu cultures. However it is still unclear as to whether arsenical bronze artefacts were imported or made locally, although the latter is suspected as being more likely due to possible local exploitation of mineral resources. On the other hand, the artefacts show typological connections to the Eurasian steppe.[*][11]

The Eneolithic period in Northern Italy, with the Remedello and Rinaldone cultures in 2800 to 2200 BCE, saw the use of arsenical bronze. Indeed, it seems that arsenical bronze was the most common alloy in use in the Mediterranean basin at this time.[*][12]

In South America, arsenical bronze was the predominant alloy in Ecuador and north and central Peru, because of the rich arsenic bearing ores present there. By contrast, the south and central Andes, southern Peru, Bolivia and parts of Argentina, were rich in the tin ore Cassiterite and thus did not use arsenical bronze.[*][7]

The Sican Culture of north western coastal Peru is famous for its use of arsenical bronze during the period 900 to 1350 AD.[*][13] Arsenical bronze co-existed with tin bronze for in the Andes, probably due to its greater ductility which meant it could be easily hammered into thin sheets which were valued in local society.[*][7]

19.4   Arsenical bronze after the Bronze Age

The archaeological record in Egypt, Peru and the Caucasus suggests that arsenical bronze was produced for a time along-side tin bronze. At Tepe Yahya its use continued into the Iron Age for the manufacture of trinkets and decorative objects,[*][3] thus demonstrating that there was not a simple succession of alloys over time, with superior new alloys replacing older ones. There are few real advantages metallurgically for the superiority of tin bronze,[*][1] and early authors suggested that arsenical bronze was phased out due to its health effects. It is more likely that it was phased out in general use because alloying with tin gave castings which had similar strength to arsenical bronze but did not require further work-hardening to achieve useful strength.[*][6] It is also probable that more certain results could be achieved with the use of tin, because it could be added directly to the copper in specific amounts, whereas the precise amount of arsenic being added was much harder to gauge due to the manufacturing process.[*][7]

19.5   Health effects of arsenical bronze use

Arsenic is an element with a vaporization point of 615°C, such that arsenical oxide will be lost from the melt before or during casting, and fumes from fire setting for mining and ore processing have long been known to attack the eyes, lungs and skin.[*][14]

Chronic arsenic poisoning leads to peripheral neuropathy, which can cause weakness in the legs and feet. It has been speculated that this lay behind the legend of lame smiths, such as the Greek god Hephaestus.

A well-preserved mummy of a man who lived around 3,200 BCE[*][15] found in the Ötztal Alps, popularly known as Ötzi, showed high levels of both copper particles and arsenic in its hair. This, along with Ötzi's copper axe blade, which is 99.7% pure copper, has led scientists to speculate that he was involved in copper smelting.[*][16]

19.6   Modern uses of arsenical bronze

Arsenical bronze has seen little use in the modern period. It appears that the closest equivalent goes by the name of arsenical copper, being defined as copper with under 0.5 wt% As, below the accepted percentage in archaeological artefacts. The presence of 0.5 wt% arsenic in copper lowers the electrical conductivity to 34% of that of pure copper, and even as little as 0.05 wt% decreases it by 15%.[*][7] Therefore, there is no demand for copper containing arsenic in electric wires etc., one of the major modern uses of copper and steam engine boilers are no longer made from it, leading to no modern use.

19.7 See also

- Arsenical copper

19.8 References

[1] Charles, J. A. (January 1967). "Early Arsenical Bronzes – A Metallurgical view". *American Journal of Archaeology* **71** (1): 21–26. JSTOR 501586.

[2] P Budd and B S Ottoway. 1995. Eneolithic Arsenical copper – chance or choice? In: Borislav Jovanovic (Ed) Ancient mining and metallurgy in southeast europe, International symposium, Archaeological institute, Belgrade and the Museum of mining and metallurgy, Bor, page 95.

[3] Thornton, C.P.; Lamberg-Karlovsky, C.C.; Liezers, M.; Young, S.M.M. (2002). "On pins and needles: tracing the evolution of copper-based alloying at Tepe Yahya, Iran, via ICP-MS analysis of Common-place items.". *Journal of Archaeological Science.* 29 If a great deal of oxygen is dissolved (29): 1451–1460. doi:10.1006/jasc.2002.0809.

[4] Lechtman, H.; Klein, S. (1999). "The Production of Copper–Arsenic Alloys (Arsenic Bronze) by cosmelting: Modern Experiment, Ancient Practice". *Journal of Archaeological Science* **26** (26): 497–526. doi:10.1006/jasc.1998.0324.

[5] De Ryck, I.; Adriens, A.; Adams, F. (2005). "An overview of Mesopotamian bronze metallurgy during the 3rd millennium BC" (PDF). *Journal of Cultural Heritage* **6** (6): 261–268. doi:10.1016/j.culher.2005.04.002.

[6] Tylecote, R.F. (1992). *A History of Metallurgy* (2nd ed.). London: Maney publishing. ISBN 0-901462-88-8.

[7] Lechtman, Heather (Winter 1996). "Arsenic Bronze: Dirty Copper or Chosen Alloy? A View from the Americas". *Journal of Field Archaeology* **23** (4): 477–514. doi:10.2307/530550. JSTOR 530550.

[8] Thornton, C.P.; Rehren, T.; Piggot, V.C. (2009). "The production of speiss (iron arsenide) during the Early Bronze Age in Iran.". *Journal of Archaeological Science* **36** (36): 308–316. doi:10.1016/j.jas.2008.09.017.

[9] Ryndina, N. 2009. The potential of metallography in investigations of early objects made of copper and copper-based alloys. Journal of the historical metallurgy society. 43, 1-18.

[10] The Nahal Mishmar Treasure at Metropolitan Museum

[11] Jianjun Mei, page 9 in Metallurgy and Civilisation, Eurasia and beyond, ed: Jianjun Mei and Thilo Rehren. Proceedings of the 6th international conference on the beginnings of the use of meals and alloys (BUMA VI), 2009,Archetype publications, London.

[12] Eaton, E. R. 1980. Early metallurgy in Italy. In: ed. W. A. Oddy, Aspects of early metallurgy, occasional paper 17, British Museum Publications, London.

[13] Hörz, G.; Kallfass, M. (December 1998). "Metalworking in Peru, ornamental objects from the Royal Tombs of Sipan". *Journal of Materials* **50** (12): 8. doi:10.1007/s11837-998-0298-2.

[14] Harper, M. (1987). "Possible toxic metal exposure of prehistoric bronze workers". *British Journal of Industrial Medicine* **44** (44): 652–656. doi:10.1136/oem.44.10.652.

[15] Age determination of tissue, bone and grass samples from Ötztal Ice Man (PDF; 476 kB)

[16] <Please add first missing authors to populate metadata.> (16 September 2002), *Iceman's final meal*, BBC News

19.9 External links

- "Arsenical copper carpenters tools from Naxos, circa 2700 to 2200 BC". British Museum.

- "Results page, with some information on arsenical bronze". Sican archaeological project.

Chapter 20

Middle Bronze Age migrations (Ancient Near East)

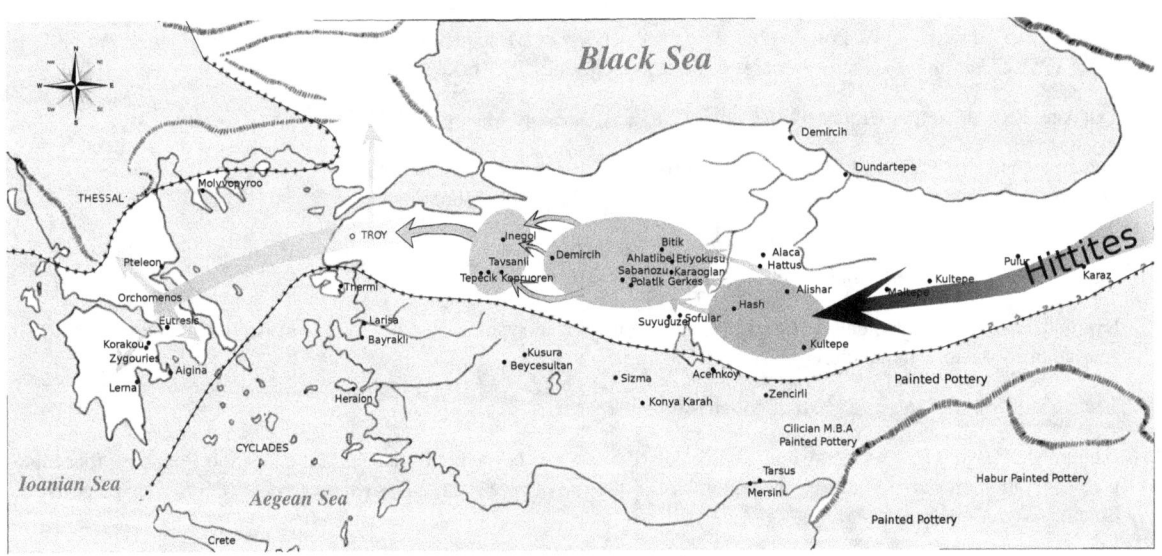

Migrations in Anatolia around 1900 BCE based on outdated research. According to Drews and Mellart, the Hittite migration displaced other peoples living in Anatolia, who in turn displaced the Middle Helladic Greek-speaking peoples to the west.[1]*[2] This is contradicted by newer research.*[3]

Various and currently outdated theories have been proposed that postulate waves of migration during the Middle Bronze Age in the Ancient Near East. While the turmoils that separate the Late Bronze Age from the Early Iron Age are well documented (see Bronze Age collapse), theories of migration during the Middle Bronze Age (20th century BCE) have little direct support.

Some suggestions connect these alleged "mass migrations" with the coming of the Greeks, moving from their former settlements into the southern and central Balkans displacing the former non-Greek inhabitants of Greece.*[1]*[4] Others make reference to a supposed migration of the Hittites to their earliest known home in Kültepe during the same period.*[2] However, newer evidence and theories contradict the notion of a migration of the Hittites, suggesting that a Proto-Indo-Hittite language dates back to the fourth or eight millennium BC.*[3]

20.1   Background

20.1.1 Hittite invasion

For reasons unknown, the Hittites moved into central Asia Minor, conquering the Hattians and later adopting their culture and name. This invasion by the Hittites displaced other peoples living in Anatolia, who in turn displaced the Middle Helladic Greek-speaking peoples to the west. This enforced an exodus from Northwestern Anatolia created a wave of refugees who invaded what is now southern Greece and destroyed the Early Helladic civilization.[*][2]

20.1.2 Destruction

Archaeological evidence shows that the cities of Erzerum, Sivas, Pulur Huyuk near Baiburt, Kultepe near Hafik, and Maltepe near Sivas were destroyed during the Middle Bronze Age. The great trading city of Kanesh (Level II) was also destroyed. From there in the hill country between Halys the destruction layers from this time tell the same story. Karaoglan, Bitik, Polatli and Gordion were burnt, as well as Etiyokusu and Cerkes. Further west near the Dardanelles the two large mounds of Korpruoren and Tavsanli, west of Kutahya, show the same signs of being destroyed.

The destruction even crossed into Europe in what is now Bulgaria. The migration brought an end to Bulgaria's Early Bronze Age, with archaeological evidence showing that the Yunacite, Salcutza, and Esero centers had a sudden mass desertion during this time.[*][2]

20.1.3 Into Greece

From the Dardanelles, the refugee invaders moved into mainland Greece, and the Peloponnese saw burnt and abandoned cities on par with the much later Dorian invasion which destroyed the Mycenaean civilization.[*][2] At this time, 1900 BC, destruction layers can be found at southern Greek sites like Orchomenos, Eutresis, Hagios Kosmas, Raphina, Apesokari, Korakou, Zygouries, Tiryns, Asine, Malthi and Asea. Many other sites are deserted, e.g. Yiriza, Synoro, Ayios Gerasimos, Kophovouni, Makrovouni, Palaiopyrgos, etc. This destruction across Greece also coincided with the arrival of a new culture that had no connection with the Early Helladic civilization, who were the original inhabitants.[*][2] Northern Greece escaped destruction, as well as southern Anatolia, which during this time showed no disturbances.[*][2]

20.1.4 Minyan ware

Main article: Minyan ware

Gray Minyan ware was first identified as the pottery introduced by this mass movement of new populations into southern Greece around 1900 BC.[*][1][*][2] However, this theory was disproved in the 1950s when excavations at Lerna showed that Minyan ware had a predecessor in the preceding Early Helladic III Tiryns culture.[*][5] The advent of Minyan ware coincides with domestic processes reflective of the smooth transition from Early to Middle Bronze Age culture.[*][6]

20.2 See also

- Bronze Age Greece

- Bronze Age Anatolia

20.3 References

20.3.1 Citations

[1] Drews 1994, p. 14.

[2] Mellaart 1958, pp. 9–23.

[3] Steadman & McMahon 2011, p. 704.

[4] Dietrich 1974, p. 4.

[5] Pullen 2008, p. 40; French 1973, pp. 51–57; Caskey 1960, pp. 285–303.

[6] Edwards, Gadd & Hammond 1971, Chapter XXIV(a) Anatolia, c. 2300–1750, p. 682: "Elsewhere the transition from Early to Middle Bronze Age culture seems to have been a smooth domestic process, unaffected by foreign influences. At individual sites, such as Troy, new wares take the place of old; but the arrival, for instance, of the so-called grey 'Minyan' pottery, which is now known to have been in use long before in neighbouring areas, suggests rather a peaceful acquisition rather than a foreign intrusion."

20.3.2 Sources

- Caskey, John L. (July–September 1960). "The Early Helladic Period in the Argolid". *Hesperia* (The American School of Classical Studies at Athens) **29** (3): 285–303. doi:10.2307/147199.

- Dietrich, Bernard Clive (1974). *The Origins of Greek Religion*. Berlin: Walter de Gruyter. ISBN 3-11-003982-6.

- Drews, Robert (1994). *The Coming of the Greeks: Indo-European Conquests in the Aegean and the Near East*. Princeton, NJ: Princeton University Press. ISBN 0-691-02951-2.

- Edwards, Iorwerth Eiddon Stephen; Gadd, C.J.; Hammond, N.G.L. (1971). *The Cambridge Ancient History (Volume II, Part I): The Early History of the Middle East*. Cambridge and New York: Cambridge University Press. ISBN 978-0-52-107791-0.

- French, D.M. (1973). "Migrations and 'Minyan' pottery in western Anatolia and the Aegean". In Crossland, R.A.; Birchall, Ann. *Bronze Age Migrations in the Aegean*. Park Ridge, NJ: Noyes Press. pp. 51–57.

- Mellaart, James (January 1958). "The End of the Early Bronze Age in Anatolia and the Aegean". *American Journal of Archaeology* **62** (1): 9–33. doi:10.2307/500459.

- Pullen, Daniel (2008). "The Early Bronze Age in Greece". In Shelmerdine, Cynthia W. *The Cambridge Companion to the Aegean Bronze Age*. Cambridge and New York: Cambridge University Press. pp. 19–46. ISBN 978-0-521-81444-7.

- Steadman, Sharon R.; McMahon, Gregory (2011). *The Oxford Handbook of Ancient Anatolia: 10,000-323 BCE*. Oxford and New York: Oxford University Press. ISBN 978-0-19-537614-2.

Chapter 21

Dover Bronze Age Boat

Dover Bronze Age boat is one of fewer than 20 Bronze Age boats so far found in Britain. It dates to 1575–1520 BCE. The boat was made using oak planks sewn together with yew lashings. This technique has a long tradition of use in British prehistory; the oldest known examples are from Ferriby in east Yorkshire. It is currently on display at Dover Museum.

21.1 Discovery and excavation

On 28 September 1992, construction workers from Norwest Holst (who were building the new A20 road link between Folkestone and Dover), working alongside archaeologists from the Canterbury Archaeological Trust, uncovered what remained of a large prehistoric boat thought to be 3,500 years old. This would place its origin around 1500 BC, in the Middle Bronze Age in England.

The boat was buried under a road and the burial site stretched out towards buildings. It was decided that it would be too dangerous to dig too near the buildings, so an unknown length of the boat has had to be left under the ground.

Previous attempts to remove such boats whole have been unsuccessful, so it was decided to cut the boat into sections and remove it and reassemble it afterwards.

After nearly a month of excavation 9.5 metres of the boat was eventually recovered. Depending on different views of the true size of the complete boat this 9.5 metres could be up to two thirds of the full size of the boat.

21.2 Oldest sea-going boat?

The Dour leads staight into the English Channel, so speculation has been made ever since its discovery about whether the Dover boat went to sea and sailed to the Continent. There is plenty of evidence that there was cross-Channel communication, but it is not known what kind of boats actually sailed across. Keith Miller, a regional achaeologist told the BBC that the older Ferriby Boats would have been used to cross the North Sea[1] and certainly the Ferriby Heritage Trust describe Ferriby Boat 3 as Europe's first known seacraft.[2] The BBC television programme *Operation Stonehenge: What Lies Beneath Pt 2*, broadcast on BBC Two in September 2014, describes the boat as seagoing and describes the tons of cargo it could have taken across the Channel. However, The Dover Museum consider that the Dover Bronze Age Boat is the oldest seagoing boat known, at only 1550 BC.[3] They are backed by the Time Team Special, broadcast in September 2014 on UK Channel 4, which stated that to be a proper sea-going, cross-channel vessel the boat would have to have the curved 'rocker' bottom and the (unproven) pointed bow that only the more modern Dover boat possesses. Confusingly, the Oakleaf reproduction of the Ferriby boats was given a pointed bow and the Ferriby boats are described as having curved rocker bottoms, which sounds much the same as the Dover boat.

Dover Bronze Age Boat at Dover Museum

Dover Bronze Age Boat at Dover Museum

21.3   Size

As part of the boat remains underground and there is no proof of the boat's overall shape and size, much speculation as to its total length and its shape has been made. The museum shows suggestions, but the boat could easily be little more than has been removed from the ground, or perhaps many metres longer.

The width of the boat is significant, being around 2 Metres wide it is much wider than dugout canoes of the time and can easily seat two people next to each other. It is wider than the Ferriby boats, for example.

21.4   Materials

The boat was constructed of oak planks, stitched together with yew withies and also fixed together with wooden wedges.[4] This makes it similar to the Ferriby boats, which are also stitched planks. It is, however, quite different to the Must Farm dugouts, which are not only dug out of one trunk, but the smaller, lightweight ones are made of lighter linden trunks.[5]

21.5   Conservation and re-assembly

Whilst in the ground the boat was significantly protected from being destroyed by waterlogging and a cover of silt which protected it from bacteria. After being removed from the ground the boat was kept in a waterlogged state at the Mary Rose Trust at Portsmouth. After a long process of preservation the boat returned to Dover Museum to be re-assembled in 1998.

21.6   Award-winning display

The boat is displayed in a glass case as the centrepiece of a whole floor in the museum devoted to archeology. With the boat itself is a modern reconstruction of a section of the boat, to assist in the visitors interpretation of the boat itself. The display won an award in 2000 for archeological display.

21.7   Reconstructions

21.7.1   Section

First, a full-size three-metre section of the boat was built, experimenting with techniques etc. This is also housed in the Dover Museum with the original.

21.7.2   Half-size reproduction

Then a half-size reconstruction of the Dover Boat was completed in Dover in 2012. Hopes to launch it at the time failed when the boat immediately shipped a lot of water, but the boat has nevertheless been touring around the Channel area in different countries and plans are well on their way to improve its seaworthyness.[6]

The boat, initially named BC 1550 has since been officially named after one of its builders, Ole Crumlin-Pederson.[7]

As of 2014, the boat has now been sailed out from Dover Harbour and was filmed for a Time Team Special for the UK Channel 4.

21.8   See also

- Ferriby Boats – comparable Bronze Age boats from Northern England

21.9   Notes

[1] http://news.bbc.co.uk/1/hi/uk/1234529.stm

[2] http://www.ferribyboats.co.uk/ retrieved 19 September 2014

[3] http://www.dovermuseum.co.uk/Bronze-Age-Boat/Bronze-Age-Boat.aspx retrieved 19 September 2014

[4] http://www.bbc.co.uk/ahistoryoftheworld/objects/KqztJvaeRsiZdlDXMAqAEA

[5] http://www.independent.co.uk/news/science/archaeology/dug-out-canoes-found-in-record-haul-in-cambridgeshire2.html

[6] http://www.canterbury.ac.uk/news/newsrelease.asp?newsPk=1967

[7] http://www.kentonline.co.uk/deal/news/bronze-age-boat-replica-fails-to-a64485/

21.10   References

- Clark, P. 2004. *The Dover Bronze Age Boat.* Swindon: English Heritage.

21.11   External links

- Dover Museum's website about the discovery, excavation, preservation and display of the boat

- Article describing story of discovery and extraction of boat in Archaeology magazine

21.12 Text and image sources, contributors, and licenses

21.12.1 Text

- **Bronze Age** *Source:* https://en.wikipedia.org/wiki/Bronze_Age?oldid=687110658 *Contributors:* AxelBoldt, Youssefsan, William Avery, Ray Van De Walker, SimonP, Zimriel, Stevertigo, Patrick, Michael Hardy, Dominus, Gdarin, Nixdorf, Collabi, Ixfd64, IZAK, Pjamescowie, Glenn, Bogdangiusca, Rossami, Nikai, Charles Matthews, Andrevan, Reddi, Jay, The Anomebot, Steinsky, DJ Clayworth, Tpbradbury, Itai, Eik-waR, Wetman, UninvitedCompany, Nufy8, Robbot, Waerth, ChrisO~enwiki, Gregors, Altenmann, Lowellian, Hadal, Wikibot, JamesMLane, DocWatson42, Jyril, Wiglaf, Tom harrison, Aphaia, Yak, Bradeos Graphon, Everyking, Chinasaur, Per Honor et Gloria, PenguiN42, Gadfium, Keith Edkins, Quadell, Beland, Iceager, 1297, Adamsan, Icairns, Sam Hocevar, Creidieki, Okapi~enwiki, Imjustmatthew, Ukexpat, Zondor, Esperant, Mike Rosoft, D6, Jayjg, Freakofnurture, Imroy, CALR, Moverton, Discospinster, ElTyrant, Rich Farmbrough, Wclark, Vsmith, Florian Blaschke, Dbachmann, Paul August, Bender235, Malkin, Eric Forste, Lycurgus, Kwamikagami, Adambro, Bobo192, Nigelj, Wisdom89, Jguk 2, Russ3Z, Chirag, Jojit fb, Flammifer, Hargrimm, Haham hanuka, Mareino, Red Winged Duck, Alansohn, Gary, Pinar, Miranche, Atlant, Gsandi, Wtmitchell, Rebroad, Garzo, Jheald, Sleigh, HenryLi, Feezo, Woohookitty, TomTheHand, Benbest, Briangotts, Pol098, JeremyA, MONGO, Paradon, John Hill, SDC, Xiong Chiamiov, Graham87, Magister Mathematicae, BD2412, RxS, Josh Parris, Rjwilmsi, Coemgenus, Саша Стефановиҳ, CristianChirita, Isaac Rabinovitch, Makaristos, InFairness, CQJ, Kalogeropoulos, Bhadani, Sydbarrett74, RJP, Nihiltres, CarolGray, RexNL, Chobot, DVdm, Voodoom, Bgwhite, Cactus.man, YurikBot, Wavelength, Jimp, Edward Wakelin, Pigman, Akamad, Gaius Cornelius, Ksyrie, Eleassar, Wimt, Thane, Anomalocaris, SEWilcoBot, Beanyk, PM Poon, Zwobot, Syrthiss, Snarius, DeadEyeArrow, Botteville, Smaines, Deville, Codrinb, Closedmouth, IdoMagal, Reyk, Cobblet, Pádraic MacUidhir, RunOrDie, Nima.nezafati, Victor falk, That Guy, From That Show!, Luk, Bigcheesegs, Treesmill, SmackBot, Amcbride, YellowMonkey, Tom Lougheed, InverseHypercube, Unyoyega, Bomac, Jagged 85, RobotJcb, Kintetsubuffalo, Flamarande, Edgar181, Alsandro, Gilliam, Hmains, Rudytan, Skizzik, Kevinalewis, Rmosler2100, Pfhreak, SMP, Hibernian, Moshe Constantine Hassan Al-Silverburg, Afasmit, DHN-bot~enwiki, Colonies Chris, Can't sleep, clown will eat me, Jahiegel, Proofreader, TheKMan, Rrburke, VMS Mosaic, Addshore, SundarBot, The tooth, Nfleming, John D. Croft, Smokefoot, Jbergquist, Just plain Bill, Flamebroil, Tankred, Jugbo, Ged UK, SashatoBot, Bcasterline, Thanatosimii, Soap, Euchiasmus, Naphureya, Jaganath, Bo99, Sir Nicholas de Mimsy-Porpington, Fig wright, DIEGO RICARDO PEREIRA, SpyMagician, Melody Concerto, JHunterJ, Mr Stephen, WaynaQhapaq, Neddyseagoon, Halaqah, BranStark, Iridescent, Craigboy, Shoeofdeath, Newone, Richard75, Courcelles, Tawkerbot2, Kevin Murray, Trevor.tombe, TarrVetus, JForget, Ohthelameness, CmdrObot, Tanthalas39, Leevanjackson, DeLarge, Godardesque, BenGriffiths, Cydebot, ArgentTurquoise, Fatemi (usupred)~enwiki, JFreeman, Chasingsol, Fuzzibloke, Doug Weller, Optimist on the run, Walgamanus, Kozuch, Omicronpersei8, Zalgo, Saintrain, Oxonhutch, Thijs!bot, Epbr123, Chloroform42, Headbomb, Marek69, Esemono, Kathovo, Itsmejudith, Echofilm, Dmitri Lytov, PER9000, Klausness, Escarbot, Mentifisto, AntiVandalBot, QuiteUnusual, Tangerines, Smartse, Modernist, Dweir, Bèrto 'd Sèra, Gökhan, Sluzzelin, JAnDbot, MER-C, CosineKitty, Dsp13, Adresia, Db099221, East718, PhilKnight, Magioladitis, Bongwarrior, VoABot II, Mrund, JNW, Ling.Nut, Antiphus, Protodruid, Ashoichet, Animum, Cmount, Philg88, Pax:Vobiscum, Patstuart, Shinnin, MartinBot, GimliDotNet, Analytikone, Phasechange, Ravichandar84, Feijuada, Wyrdlight, R'n'B, Mycroft7, CommonsDelinker, Puff Of Hot Air, Nono64, Jarhed, J.delanoy, Bogey97, Maurice Carbonaro, Acalamari, Katalaveno, Johnbod, McSly, Bayinnaung, Balthazarduju, Dmitri Yuriev, AntiSpamBot, NewEnglandYankee, ThinkBlue, Student7, KylieTastic, Rtelkin, Skryinv~enwiki, Jevansen, DorganBot, Yellowfiver, Idioma-bot, Markrimmell, Ariobarzan, Malik Shabazz, VolkovBot, DSRH, FriendlyDalek, AlnoktaBOT, Historiographer, Achim Jäger~enwiki, Soliloquial, Super Knuckles, Closms, TXiKiBoT, Mercurywoodrose, Laughingyet, Wiikipedian, Martin451, JhsBot, Don4of4, Jackfork, Wikiisawesome, Luuva, Kiinslayer, Emma Ephemera, JemGage, Falcon8765, Spinningspark, Bryncaderfaner, Spitfire8520, Mike4ty4, AlleborgoBot, Ekendrick, PericlesofAthens, EmxBot, SieBot, Winchelsea, Gerakibot, Calabraxthis, Keilana, Tiptoety, Oda Mari, Hxhbot, Carnun, Yerpo, Hzh, Tombomp, WSBaird, MagnusF, OKBot, Dcattell, Superbeecat, Illinois2011, Pinkadelica, Brenny91, Velvetron, ShajiA, Explicit, Missing Ace, Atif.t2, Martarius, Apuldram, ClueBot, The Thing That Should Not Be, Wolfch, Drmies, CounterVandalismBot, Niceguyedc, The Wild West guy, Nikzbitz, Jos.knight, Parkwells, Neverquick, Puchiko, Chris Kutler, Alex836, Grunty Thraveswain, DragonBot, Hustead, DavidBlackwell, Excirial, AssegaiAli, Quercus basaseachicensis, Canis Lupus, Jusdafax, Kanguole, Jacksinterweb, Jayantanth, Tnxman307, BOTarate, Calor, Cold Phoenix, Thingg, Notthe600, Mamenchisaurus, Versus22, DumZiBoT, SilvonenBot, Vegas949, Tkech, Ploversegg, Leonid111, Jadtnr1, Rubberbiscuits, Luwilt, Addbot, Roentgenium111, Some jerk on the Internet, ImPods, Ronhjones, TutterMouse, Zarcadia, CanadianLinuxUser, Doremon360, Diptanshu.D, Swiveler, Proxima Centauri, Zymethdemon, Den123456dk, Numbo3-bot, Ehrenkater, Tide rolls, Lightbot, Krano, Jarble, Margin1522, Legobot, Luckas-bot, Yobot, DerechoReguerraz, Senator Palpatine, Amirobot, Rud Hud Hudibras, Becky Sayles, Getoryk, AnakngAraw, EddyBSt, AnomieBOT, Andrewrp, 1exec1, Captain Quirk, Jim1138, Kingpin13, ImmigrantUS, Materialscientist, Sea888, Viletraveller, Maxis ftw, Frankenpuppy, ArthurBot, Madalibi, Xqbot, Timir2, Capricorn42, GenQuest, Hanberke, Tad Lincoln, Br77rino, Aa77zz, Crzer07, Petropoxy (Lithoderm Proxy), LevenBoy, J04n, GrouchoBot, RibotBOT, Reginald Molehusband, GhalyBot, Tulocci, Aachen123, Shadowjams, Kickyandfun, E0steven, FrescoBot, Cunibertus, Anna Roy, Finalius, Killermanross, Stephenmiller10, Citation bot 1, Þjóðólfr, Pinethicket, I dream of horses, Tsingee, RedBot, Hantzen, Froaringus, Koakhtzvigad, FoxBot, TrickyM, TobeBot, Zoeperkoe, Vrenator, Tenea32, Kmk1997, Seshnakht, Tbhotch, Minimac, Mean as custard, RjwilmsiBot, BonGrok, Slon02, DASHBot, EmausBot, John of Reading, Milkunderwood, AGStewart, Lboscher, Lynx82, Liambr101, GoingBatty, Slightsmile, Wikipelli, AvicBot, Vanished user sdjei4o346jowe3, Érico Júnior Wouters, Oncenawhile, Wangnan113, Confession0791, Wayne Slam, Donner60, Bad edits r dumb, Joshblood, Rememberway, ClueBot NG, Gothicartech, Znoes, MelbourneStar, Chester Markel, Dalesti, Shocks30, Widr, PatHadley, Highspeedrailguy, Helpful Pixie Bot, Titodutta, Gob Lofa, BG19bot, Chelsealb, Vagobot, Hallows AG, MusikAnimal, AvocatoBot, Amp71, Mark Arsten, Compfreak7, Iamthecheese44, FoxCE, SmellyLilYou, Joshua Jonathan, Romulan Ale, Glevum, EJcarter, Zedshort, Gibbja, BattyBot, R3venans, DarafshBot, Derslek, Dexbot, Crumpled Fire, Hmainsbot1, Mogism, Qazzaqazzaq, Sheldon65, Lugia2453, Frosty, Hair, WilliamDigiCol, Krakkos, Foonarres, Corinne, Epicgenius, Yulin Zhang, Vanamonde93, Dustin V. S., DavidLeighEllis, Monochrome Monitor, S-adchen, JustBerry, Werddemer, JERIN.L.VARGHESE, Afro-Eurasian, Stamptrader, Cencer90, Epicface124, Cancina5645, Artist robert jones, Prkprescott, Monkbot, Attar-Aram syria, BlueAlps, Braden12345, Crom daba, Rsuwsearch, Johnhart151, Z. PUPU, Liann2009, Piesquared93, Cookingtheworld, Asdklf;, Calamity Fortune, Seapre, Julietdeltalima, Nadavisediter3345, Olitrampoline, StewdioMACK, WyattAlex, Grathmy, Niall Williams, KasparBot, Srednuas Lenoroc, BU Rob13, O3fkz, Slimy goop, Cool123zs and Anonymous: 680

- **Three-age system** *Source:* https://en.wikipedia.org/wiki/Three-age_system?oldid=684284004 *Contributors:* Bryan Derksen, Michael Hardy, Nixdorf, Sannse, Julesd, Smack, Reddi, Peregrine981, Itai, Penfold, Rogper~enwiki, Sam Spade, Lowellian, Dbenbenn, Confuzion, Beland,

sui, Bobo192, Longhair, Robotje, Smalljim, Enric Naval, JW1805, Juzeris, Darwinek, PWilkinson, Idleguy, Krellis, Ranveig, Jumbuck, Alansohn, Gary, Eric Kvaalen, Arthena, Wiki-uk, Keenan Pepper, Riana, LRBurdak, Kurt Shaped Box, Lightdarkness, Arunreginald, Snowolf, Aranae, Velella, Amorymeltzer, Someoneinmyheadbutit'snotme, HGB, Oleg Alexandrov, TShilo12, Anthony aragorn, Sjv27~enwiki, OwenX, Woohookitty, Ghane, Mr Tan, Camw, MONGO, Schzmo, Dangerous-Boy, GregorB, John Hill, Hughcharlesparker, Wayward, Palica, Matilda, Raguks, Graham87, BD2412, Kbdank71, Dpv, Ahsen, Rjwilmsi, Erebus555, Ikh, Durin, Karma Thief, Nguyen Thanh Quang, Vuong Ngan Ha, FlaBot, Mishuletz, AdnanSa, Darcyj, Margosbot~enwiki, Nivix, RexNL, ChongDae, CJLL Wright, Chobot, DaGizza, Moocha, DVdm, Bgwhite, Gwernol, Roboto de Ajvol, YurikBot, Wavelength, Deeptrivia, Phantomsteve, RussBot, Gaius Cornelius, Vyzasatya, Rsrikanth05, Srini81, NawlinWiki, Kantokano, Grafen, Siddiqui, Joel7687, Stallions2010, Romarin, Bota47, Mjsabby, Szhaider, Ms2ger, AjaxSmack, Crisco 1492, Boivie, Rudrasharman, Closedmouth, Ketsuekigata, Scriber~enwiki, Extraordinary, Will R Turner, Xaxafrad, Pratheepps, WIN, Katieh5584, That Guy, From That Show!, SpLoT, Sardanaphalus, SmackBot, Unschool, David Kernow, Yusufkhan, KnowledgeOfSelf, Waqas.usman, Pgk, Korossyl, Jagged 85, Tarzenda, Frymaster, Kintetsubuffalo, Aivazovsky, Gilliam, Hmains, Skizzik, Chris the speller, Bluebot, Ian13, Sandycx, CSWarren, Viewfinder, Baronnet, Hongooi, Darth Panda, Rama's Arrow, Can't sleep, clown will eat me, DéRahier, JRPG, OrphanBot, Rrburke, Krsont, Addshore, Abulfazl, SundarBot, Cribananda, Decltype, Zdravko mk, The unbearable brightness of beam, Nmpenguin, Shahkhan, Bejnar, Drunken Pirate, Andrew Dalby, Ohconfucius, Apalaria, Mircea, Unre4L, Carnby, Kipala, Benesch, Shyamsunder, Green Giant, Aarandir, Ckatz, JHunterJ, Koroviaff, Luokehao, Meco, Deepak D'Souza, MTSbot~enwiki, Skinsmoke, Hu12, BranStark, Spartian, Shoeofdeath, Geoffg, Ehsankiani, Courcelles, Tawkerbot2, RaviC, Ghaly, Thetrick, SkyWalker, Sooku, Mattbr, Primeboy, JohnCD, Ibadibam, Neelix, Myasuda, Equendil, Cydebot, Ramitmahajan, Gogo Dodo, Hebrides, Vargob, Ferretsnarf, Legacystrike, Tawkerbot4, JodyB, Mattisse, Thijs!bot, Epbr123, Islescape, Sarfraz777, Dpall, Anupam, Marek69, Esowteric, OrenBochman, Nick Number, Escarbot, Pie Man 360, AntiVandalBot, Luna Santin, Seaphoto, Robzz, Peterwinn, Tigeroo, Sluzzelin, JAnDbot, Samar, Ekabhishek, Bakasuprman, MER-C, Attarparn, Magioladitis, Bongwarrior, VoABot II, AuburnPilot, AtticusX, Khalidkhoso, CTF83!, Tinucherian, Sindhutvavadin, Alishehzad, Prester John, JaGa, Philg88, Khalid Mahmood, BeckyLadakh, Pax:Vobiscum, Hbent, Rickard Vogelberg, Afil, S3000, Atulsnischal, Jackson Peebles, Salvager, SI4, Cody6, Kevinsam, Nkadambi, Rettetast, R'n'B, Sikh-history, AlexiusHoratius, Nono64, AgarwalSumeet, LedgendGamer, Wlodzimierz, J.delanoy, Pharaoh of the Wizards, Trusilver, Asif110, Eliz81, PC78, Lantonov, Katalaveno, Juliancolton, DH85868993, Nomi887, CardinalDan, Idioma-bot, Pclift, Redtigerxyz, VolkovBot, Haim Berman, Jeff G., Indubitably, KindGoat, Abid Khan Seengharay (Yousafzai), Philip Trueman, TXiKiBoT, Pahari Sahib, Suprah, Sudipta.dasgupta, Uch, Martin451, Sniperz11, LeaveSleaves, Andrewrost3241981, Natg 19, ARUNKUMAR P.R, Azhar aslam, Kmhkmh, Kotai~enwiki, Adam.J.W.C., Falcon8765, Turgan, Onceonthisisland, AlleborgoBot, Tvinh, Roland zh, Junoon53, Vsst, EmxBot, Arjun024, Sojharo, SieBot, Zip600001, Tiddly Tom, Lemonflash, Smsarmad, Bentogoa, Flyer22 Reborn, SPACKlick, CutOffTies, Oxymoron83, Antonio Lopez, Vmrgrsergr, Fratrep, Adam Cuerden, Jacob.jose, Motyka, ImageRemovalBot, Atif.t2, Apuldram, ClueBot, UrsusArctosL71, Tanmay110, The Thing That Should Not Be, Matdrodes, Saddhiyama, Khanpride, Niceguyedc, Raawais, Arunsingh16, Siyal1990, Jfdavis668, Xinjao, DragonBot, Takeaway, Jusdafax, PixelBot, Gtstricky, Wiki dr mahmad, Rao Ravindra, World, Jotterbot, Koppoliy, SchreiberBike, BOTarate, Chaosdruid, Thingg, Mhockey, Burner0718, Apparition11, Ashixamo, XLinkBot, Roxy the dog, Drm 1976, Dthomsen8, P30Carl, Mfarooqtariq, Mm40, Jbeans, Seneca13, Bgag, Addbot, Shree90, Misaq Rabab, Ronhjones, Njaelkies Lea, Cuaxdon, Fluffernutter, NjardarBot, Ka Faraq Gatri, Talha, AndersBot, Favonian, West.andrew.g, Kahasabha, Numbo3-bot, Paknur, Issyl0, Tide rolls, Lightbot, Geronimo81, Gail, Ben Pirard, Legobot, Luckas-bot, Shannon1, Yobot, 2D, Now05ster, Siddharthssj4, Pectore, R.steenhard, Ata Fida Aziz, KamikazeBot, Pk5abi, 臨海, AnomieBOT, Neptune5000, Ulric1313, Materialscientist, Maxis ftw, Thecalculator98, Rockoprem, RajeshUnuppally, Xqbot, TinucherianBot II, Ali944rana, Gigemag76, Johnxxx9, Almabot, J04n, Doorvery far, Armbrust, Nayvik, Omnipaedista, RibotBOT, SassoBot, The Interior, Brickline, Chandan Guha, Verbum Veritas, Shadowjams, AlexanderVanLoon, Rahul1365, Nagualdesign, FrescoBot, Originalwana, Gbondmg42, Massagetae, Teckgeek, Hussain Ahmad Faizy, K.Khokhar, Ahmer Jamil Khan, Lilaac, Citation bot 1, Taeyebaar, Pinethicket, I dream of horses, Rayshade, HRoestBot, Rushbugled13, King Zebu, Hamtechperson, Kraj35, Golden Penguin, FoxBot, TobeBot, Lotje, Fox Wilson, Harut8, Irfannaseefp, Diannaa, Pownerus, Akgravgaard, Brumon, DARTH SIDIOUS 2, User Team, Policysukz, DRAGON BOOSTER, CalicoCatLover, DASHBot, EmausBot, Mzr20, Dadaist6174, Katherine, Racerx11, Junebug696, Toutvientapoint, Khalid69, Tommy2010, Sheeana, Djembayz, Mekong2, AvicBot, HiW-Bot, ZéroBot, Mar4d, Bm1996, Khaqanamin, Makecat, Mdmday, Sindhulogy, Rajadevjani, Tolly4bolly, Cyberdog958, L Kensington, Lesto101, Donner60, Drustaz, ChuispastonBot, NTox, Мурад 97, TYelliot, DASHBotAV, Rocketrod1960, Khestwol, ResearchRave, ClueBot NG, Jack Greenmaven, Ramtejvarma, Powstini, LogX, Adair2324, Teepusultan, Bolori, Babanwalia, Prabhat1729, Helpful Pixie Bot, Ani23390, Khahori, Woodszack, Vagobot, PhnomPencil, Mysterytrey, Uhlan, Frze, Jogi don, Onewhohelps, Jeena1986, Carlstak, Tangerinehistry, Jayadevp13, Tanu21, Lieutenant of Melkor, Khahori01, Kriteesh, Dav subrajathan.357, Jeremy112233, Athira Rajkamal, Pratyya Ghosh, Kalyanisuresh, Mdann52, Mrt3366, Cyberbot II, Farvartish, GoShow, AzseicsoK, NitRav, Propaganda Charisma, JYBot, EagerToddler39, BrightStarSky, Sminthopsis84, Souparnikaachu1, Kiranpaul143, Delljvc, Lugia2453, Frosty, SFK2, Ulfrik Stormcloak, Cptcha, WBRSin, Joemanderson75, Scoodlypoopin, Obaid Raza, Hillbillyholiday, Nirmaladvani, Billy Boy 69, Raviteja338, HistoryofIran, Ribena786, Hoppeduppeanut, Guykom, Flat Out, LouisAragon, Babitaarora, Sushilmishra, Coolgama, Saladin1987, Ugog Nizdast, Nikhilmn2002, Kevinsydong1, Visakha veera, Bladesmulti, PJDF2367, Ashyboy67, Ezequial mendoza3, Monkbot, Nestwiki, Prymshbmg, Rd1walker, BethNaught, Thundermann, Shuaib Qureshi, UsmanKhan, Hijigne, ReneVermeulen, Chaitanyaanand1, A.stationary.traveller, Sherlybobs, Rawfey, Alicia Florrick, Nishtiaqrana, BodduLokesh, Morningwood4, Arvansages, D= Im lonely =D, SamaaNews, Power22, Human3015, Randhwasingh, Conradjagan, Weirdo103, Ankush 89, KasparBot, Aizaz Nabi Bhatti, Mascot2244, Deepanshu1707, Filpro and Anonymous: 812

- **Prehistory of Anatolia** *Source:* https://en.wikipedia.org/wiki/Prehistory_of_Anatolia?oldid=686411870 *Contributors:* Tourguide, Bender235, Kwamikagami, CeeGee, Tabletop, Bgwhite, Botteville, Hmains, Hibernian, HJJHolm, Michael Goodyear, EtienneDolet, R'n'B, CommonsDelinker, DrKay, Shinju, Denisarona, Wikihistorian, Arjayay, SchreiberBike, Bgag, Yobot, Alexikoua, Madalibi, January2009, FrescoBot, JL 09, Aamsse, Zoeperkoe, Pandukht, ClueBot NG, Frietjes, Wllmevans, Helpful Pixie Bot, RudolfRed, BattyBot, Hmainsbot1, Cavann, WillemBK, Kaputak, L0st H0r!z0ns and Anonymous: 11

- **History of the ancient Levant** *Source:* https://en.wikipedia.org/wiki/History_of_the_ancient_Levant?oldid=672897463 *Contributors:* Kpjas, Marj Tiefert, Derek Ross, Bryan Derksen, MarXidad, DanKeshet, Andre Engels, LA2, Jkominek, Olivier, Rickyrab, Stevertigo, Lir, JohnOwens, Llywrch, MartinHarper, IZAK, Sannse, Paul A, Egil, Francs2000, Chris Roy, Flauto Dolce, Humus sapiens, PBP, GreatWhiteNortherner, Snobot, Marcika, Yak, Junkyardprince, Adamsan, D6, Rich Farmbrough, Dbachmann, Grutter, El C, Vanished user kjij32ro9j4tkse, Bobo192, Foobaz, Jmdavid, Helix84, Pouya, Bart133, Jheald, Tainter, Woohookitty, G.W., Graham87, Dpv, Gramaic, Yuber, Ian Pitchford, YurikBot, NTBot~enwiki, SingingDragon, RussBot, Gaius Cornelius, Rob117, JLaTondre, Cotoco, Mmcannis, SmackBot, PiCo, Elonka, Federalist51, Hmains, Jprg1966, Sambazzi, Tewfik, Cplakidas, Drsmoo, Mr.Z-man, Khoikhoi, Radagast83, John D. Croft, Dreadstar, Joseph

Codrinb, Arthur Rubin, Saukkomies, Ajdebre, SmackBot, Arny, Hmains, Hibernian, Future Perfect at Sunrise, Marek69, Dmitri Lytov, Dc76, R'n'B, CommonsDelinker, Konkorde2, Meiskam, Gogmourn, Strich3d, Thanatos666, Slovenski Volk, Angelo De La Paz, 3rdAlcove, Squash Racket, ImageRemovalBot, Athenean, Kathleen.wright5, Boing! said Zebedee, SchreiberBike, Catalographer, Borsoka, Emmette Hernandez Coleman, Addbot, Bratislav, LaaknorBot, Ben Ben, Yobot, Againme, Alexikoua, RaulOancea, Omnipaedista, Mjasfca, Julia Clifton, FrescoBot, Olovni, MGA73bot, Dogaru Florin, Vinie007, Dewritech, Tomobe03, ClueBot NG, Hazbulator, BaboneCar, Ecad93, Dream of Nyx, WIlmevans, PatHadley, Jeremy112233, YFdyh-bot, Khazar2, Hmainsbot1, Cavann and Anonymous: 33

- **Bronze Age in Romania** *Source:* https://en.wikipedia.org/wiki/Bronze_Age_in_Romania?oldid=681456490 *Contributors:* Gadget850, Codrinb, Derek R Bullamore, Funnyfarmofdoom, Biruitorul, CommonsDelinker, Hugo999, Shinju, Addbot, Ironholds, Yobot, Bunnyhop11, Sarrus, FrescoBot, HRoestBot, Tea with toast, AvicBot, HiW-Bot, Mateicovrig, BG19bot, Ramesh Ramaiah, GreenGibbon, Leefkrust22, Mogism, VoxelBot, ProjektShir DAI, Philosopher kat, Afro-Eurasian, Jashiwat, TeiloTrimble and Anonymous: 2

- **Atlantic Bronze Age** *Source:* https://en.wikipedia.org/wiki/Atlantic_Bronze_Age?oldid=686082439 *Contributors:* Michael Hardy, Genie, Reddi, Fergananim, Confuzion, Florian Blaschke, Dbachmann, Bender235, Ricky81682, Sugaar, Rjwilmsi, Sceptre, RussBot, Pigman, Theelf29, The Ogre, Botteville, SmackBot, TharkunColl, Hmains, Cattus, Hibernian, Paul S, Cydebot, Dmitri Lytov, Antiphus, Skumarlabot, IceDragon64, HighKing, Black Kite, Raggz, 3rdAlcove, Chronicler~enwiki, Addbot, Luckas-bot, Yobot, MidnightBlueMan, Joostik, Archaeodontosaurus, Actarus Prince d'Euphor, RedBot, Froaringus, Trappist the monk, Lotje, Böri, RjwilmsiBot, DASHBot, Vanished user sdjei4o346jowe3, Midas02, Bear32ie, Gob Lofa, Gallaecian, Askatuga and Anonymous: 9

- **Bronze Age Britain** *Source:* https://en.wikipedia.org/wiki/Bronze_Age_Britain?oldid=686133529 *Contributors:* Reddi, Fergananim, Florian Blaschke, Dbachmann, Nigelj, Wtmitchell, Woohookitty, Old Moonraker, John Maynard Friedman, Botteville, SmackBot, Hmains, Chris the speller, Hibernian, Balin42632003, Midnightblueowl, Lottamiata, Mike Christie, Doug Weller, Dmitri Lytov, PhilKnight, Antiphus, J.delanoy, It Is Me Here, Johnbod, NewEnglandYankee, IceDragon64, Gpier66, Mudwater, Motacilla, Ekendrick, Francish7, Addbot, Ehrenkater, Yobot, Mattis, FrescoBot, RjwilmsiBot, Vanished user sdjei4o346jowe3, ClueBot NG, Hawa-Ave, Frietjes, Sigwells, Lincoln Josh, Helpful Pixie Bot, Gob Lofa, Biere D'or, HMSLavender and Anonymous: 13

- **Nordic Bronze Age** *Source:* https://en.wikipedia.org/wiki/Nordic_Bronze_Age?oldid=685615754 *Contributors:* Nixdorf, Genie, Reddi, Haukurth, AnonMoos, Wetman, Tim Ivorson, Rursus, Mushroom, VanishedUser kfljdfjsg33k, Wiglaf, Mboverload, Adamsan, Sam Hocevar, Jkl, Dis-cospinster, Dbachmann, Longhair, SpaceMonkey, Caeruleancentaur, Patsw, Alansohn, Ricky81682, DreamGuy, Gene Nygaard, PoccilScript, Bullenwächter, Fred J, FreplySpang, Chobot, Pigman, Wiki alf, Bloodofox, Botteville, Deville, Orcaborealis, Curpsbot-unicodify, SmackBot, CrypticBacon, Peter Isotalo, Angelbo, Hibernian, OrphanBot, Stevenmitchell, Alseeger, CmdrObot, Kallerdis, Thijs!bot, Marek69, DmitriLytov, AntiVandalBot, DagosNavy, MER-C, Mrund, Berig, Nwcasebolt, J.delanoy, Weissmann~enwiki, Twinchester, Idioma-bot, Rokus01, Michael riber jorgensen, Varoon Arya, Hxhbot, Le Pied-bot~enwiki, ImageRemovalBot, Arakunem, Drmies, Cpt. Grammar, Alexbot, Eeek-ster, Fastily, Addbot, CactusWriter, Lightbot, Againme, Richigi, AnomieBOT, LilHelpa, Xqbot, SpaceTravellor, Hauganm, PHansen, Jonkerz, ClueBot NG, PatHadley, Helpful Pixie Bot, Pohjannaula, BredyLawson, Wheeke, Krakkos, RhinoMind, Mikel.j.b, KasparBot and Anonymous:47

- **Arsenical bronze** *Source:* https://en.wikipedia.org/wiki/Arsenical_bronze?oldid=687126162 *Contributors:* William Avery, Stone, Alan Liefting, Florian Blaschke, Svartalf, Xiong Chiamiov, BD2412, Rjwilmsi, Quiddity, Alphachimp, Hairy Dude, Hellbus, NawlinWiki, Victor falk, Sbosman, SmackBot, GwydionM, Thumperward, Wizard191, Iridescent, ShakespeareFan00, Headbomb, Monol~enwiki, Dekimasu, Domusaurea, Globbet, Johnbod, RobertMadugba, IceDragon64, Andy Dingley, Kpschoedel, Addbot, Jtlessl, Ettrig, Yobot, Freikorp, Citation bot, Marta Sabando, Westerncenter, Carlog3, Fortdj33, Citation bot 1, RjwilmsiBot, Hirsutism, AGStewart, Y-barton, Helpful Pixie Bot, Arminden, CitationCleanerBot, BattyBot and Anonymous: 16

- **Middle Bronze Age migrations (Ancient Near East)** *Source:* https://en.wikipedia.org/wiki/Middle_Bronze_Age_migrations_(Ancient_Near_East)?oldid=671036058 *Contributors:* Dbachmann, Art LaPella, CeeGee, Circeus, Wtmitchell, Drbreznjev, Rjwilmsi, SmackBot, Kimon, Hibernian, Egsan Bacon, Gdeyoe, Causantin, Future Perfect at Sunrise, Esemono, Magioladitis, Mrund, ClueBot, Piledhigheranddeeper, Catalographer, Addbot, Alandeus, AnomieBOT, Alexikoua, WebCiteBOT, Citation bot 1, Bobta109, Pianoplonkers, Moswento, SporkBot, Helpful Pixie Bot, ChrisGualtieri, Mogism, Krakkos, Cavann, Monkbot and Anonymous: 12

- **Dover Bronze Age Boat** *Source:* https://en.wikipedia.org/wiki/Dover_Bronze_Age_Boat?oldid=680951431 *Contributors:* Deb, Mervyn, Geos, Thefuguestate, Antriver, Alaibot, DavidShaw, Keith D, IceDragon64, Phasler90, Flyer22 Reborn, ImageRemovalBot, Thgoiter, Sun Creator, Mhockey, Addbot, Yobot, Omnipaedista, Full-date unlinking bot, Jesse V., Fenlandier, BG19bot, Foonarres, Ilovemysister777 and Anonymous: 13

21.12.2 Images

- **File:13-Urartu-9-6mta.gif** *Source:* https://upload.wikimedia.org/wikipedia/commons/6/60/13-Urartu-9-6mta.gif *License:* CC-BY-SA-3.0 *Contributors:* www.armenica.org *Original artist:* Original uploader was Artaxiad at en.wikipedia

- **File:14_century_BC_Eastern.png** *Source:* https://upload.wikimedia.org/wikipedia/commons/9/98/14_century_BC_Eastern.png *License:* CC BY-SA 3.0 *Contributors:* Own work, data taken from: *History Year by Year*, Dorling Kindersley Ltd, 2011, pages: 32-33, ISBN 1405391057, 9781405391054. Topography taken from DEMIS Mapserver, which are public domain, other wise self-made. *Original artist:* Alexikoua,

- **File:Age-of-Man-wiki.jpg** *Source:* https://upload.wikimedia.org/wikipedia/commons/3/30/Age-of-Man-wiki.jpg *License:* Public domain *Contributors:* Transferred from en.wikipedia to Commons by User:Legoktm using CommonsHelper. *Original artist:* Ernst Haeckel

- **File:Aiud_History_Museum_2011_-_Cotofeni_Culture_Pottery.JPG** *Source:* https://upload.wikimedia.org/wikipedia/commons/9/9f/History_Museum_2011_-_Cotofeni_Culture_Pottery.JPG *License:* CC BY-SA 3.0 *Contributors:* Own work *Original artist:* Codrin.B

- **File:Alba_Iulia_National_Museum_of_the_Union_2011_-_Late_Bronze_Age_Vessels_and_Bronze_Objects.JPG** *Source:* https://wikimedia.org/wikipedia/commons/5/5e/Alba_Iulia_National_Museum_of_the_Union_2011_-_Late_Bronze_Age_Vessels_and_Bronze_Objects.JPG *License:* CC BY-SA 3.0 *Contributors:* Own work *Original artist:* Codrin.B

21.12.3 Content license